Andrzej Bartoszewicz,
Aleksandra Nowacka-Leverton

Time-Varying Sliding Modes for Second and Third Order Systems

Series Advisory Board

P. Fleming, P. Kokotovic,
A.B. Kurzhanski, H. Kwakernaak,
A. Rantzer, J.N. Tsitsiklis

Authors

Prof. Andrzej Bartoszewicz

Institute of Automatic Control
Technical University of Łódź
18/22 Stefanowskiego St.
90-924 Łódź
Poland
E-Mail: andrzej.bartoszewicz@p.lodz.pl

Dr. Aleksandra Nowacka-Leverton

Institute of Automatic Control
Technical University of Łódź
18/22 Stefanowskiego St.
90-924 Łódź
Poland
E-Mail: ola.nowacka@gmail.com

ISBN 978-3-540-92216-2   e-ISBN 978-3-540-92217-9

DOI 10.1007/978-3-540-92217-9

Lecture Notes in Control and Information Sciences   ISSN 0170-8643

Library of Congress Control Number: 2008910998

© 2009 Springer-Verlag Berlin Heidelberg

This work is subject to copyright. All rights are reserved, whether the whole or part of the material is concerned, specifically the rights of translation, reprinting, reuse of illustrations, recitation, broadcasting, reproduction on microfilm or in any other way, and storage in data banks. Duplication of this publication or parts thereof is permitted only under the provisions of the German Copyright Law of September 9, 1965, in its current version, and permission for use must always be obtained from Springer. Violations are liable for prosecution under the German Copyright Law.

The use of general descriptive names, registered names, trademarks, etc. in this publication does not imply, even in the absence of a specific statement, that such names are exempt from the relevant protective laws and regulations and therefore free for general use.

Typeset & Cover Design: Scientific Publishing Services Pvt. Ltd., Chennai, India.

Printed in acid-free paper

5 4 3 2 1 0

springer.com

Lecture Notes
in Control and Information Sciences 382

Editors: M. Thoma, F. Allgöwer, M. Morari

Acknowledgments

Our work presented in this book was developed with the kind support of the Polish State budget provided in the years 2008 – 2010, under the research project "Design of the switching surfaces for the sliding mode control of dynamic plants" (N N514 300035). This work was also possible due to the generous help of the Foundation for Polish Science (FNP). The financial assistance of the Foundation, given to the second author of this monograph, Aleksandra Nowacka-Leverton under the Young Researcher Support program, is gratefully acknowledged. Furthermore, a few projects sponsored by the Polish State Committee for Scientific Research and Polish Ministry of Science and Higher Education in the past 12 years, also helped initiate and – to some extent – develop many ideas reported in this book.

The research reported in this monograph has benefited greatly from collaboration with a number of PhD students and many friends at Institute of Automatic Control, Technical University of Łódź, Poland. Therefore, we wish to thank all of them for their encouragement, cooperation and stimulating discussions while we were preparing this book. In particular, we wish to thank Prof. Piotr Ostalczyk for his comments on essential parts of the book, and Ms. Justyna Żuk for tireless proofreading of the many drafts of this book. Her exceptional proofreading skills have been of great importance for the final form of this book. We also wish to thank all those individuals at Department of Electrical, Electronic, Control and Computer Engineering at Technical University of Łódź, Poland, who gave us the opportunity to undertake and continue the research presented in this book. We acknowledge the much needed, constructive help of Prof. Zbigniew Nowacki and Dr. Paweł Adamczyk who helped us in many ways during the last couple of months when we were working on this book. We must also thank our families for their patience and continuous moral support during the period when this monograph was written. We are particularly grateful to Mr. Adam Leverton for his valuable assistance on numerous occasions and linguistic advice on the text.

Finally, we wish to thank the entire team of Springer publications for allowing the preparation of this book to proceed. We hope that the result of our joint effort will be of true interest to the control community working on various aspects of nonlinear control systems, and in particular those working in the variable structure systems community.

Andrzej Bartoszewicz
Aleksandra Nowacka-Leverton
Institute of Automatic Control,
Technical University of Łódź, Poland

Contents

1 Introduction **1**
  1.1 Preliminaries 1
    1.1.1 Variable Structure Systems 1
    1.1.2 Sliding Mode Control 4
  1.2 Previous Work on Sliding Mode Control with Time-Varying
    Switching Surfaces 10

2 Time-Varying Sliding Modes for the Second Order Systems **17**
  2.1 Control Strategy 18
  2.2 Switching Line Design Minimising IAE 23
    2.2.1 Switching Line Design Subject to Input Signal
        Constraint 24
    2.2.2 Switching Line Design Subject to Velocity Constraint 33
    2.2.3 Switching Line Design Subject to Input Signal and
        Velocity Constraints 42
  2.3 Switching Line Design Minimising ITAE 46
    2.3.1 Switching Line Design Subject to Input Signal
        Constraint 47
    2.3.2 Switching Line Design Subject to Velocity Constraint 56
    2.3.3 Switching Line Design Subject to Input Signal and
        Velocity Constraints 62

3 Time-Varying Sliding Modes for the Third Order Systems **67**
  3.1 Control Strategy 68
  3.2 Switching Plane Design Minimising IAE 72
    3.2.1 Switching Plane Design Subject to Input Signal
        Constraint 73
    3.2.2 Switching Plane Design Subject to Acceleration
        Constraint 86
    3.2.3 Switching Plane Design Subject to Velocity Constraint 99
    3.2.4 Switching Plane Design Subject to Acceleration and
        Velocity Constraints 107
    3.2.5 Switching Plane Design Subject to Acceleration and Input
        Signal Constraints 114
    3.2.6 Switching Plane Design Subject to Input Signal and
        Velocity Constraints 117

VIII   Contents

        3.2.7 Switching Plane Design with Acceleration, Velocity and
Input Signal Constraints.. 122
  3.3 Switching Plane Design Minimising ITAE 139
        3.3.1 Switching Plane Design Subject to Input Signal
Constraint... 140
        3.3.2 Switching Plane Design Subject to Acceleration
Constraint... 149
        3.3.3 Switching Plane Design Subject to Velocity Constraint..... 156
        3.3.4 Switching Plane Design Subject to Acceleration and
Velocity Constraints.. 161
        3.3.5 Switching Plane Design Subject to Acceleration and Input
Signal Constraints... 166
        3.3.6 Switching Plane Design Subject to Input Signal and
Velocity Constraints.. 169
        3.3.7 Switching Plane Design with Acceleration, Velocity and
Input Signal Constraints..................................... 173

4 Conclusions... **181**

   References... **183**

   Index... **191**

List of Symbols

A, B, c, c_1, c_2 – switching line or switching plane parameters
$A_{opt}, B_{opt}, c_{opt}, c_{1\,opt}, c_{2\,opt}$ – optimal values of parameters A, B, c, c_1, c_2
a_{max} – maximum admissible value of the system acceleration
b – function of the time and the system state
d – external disturbance
$\boldsymbol{e}$ – tracking error vector
e_1, e_2, e_3 – tracking error vector components
f – function of the time and the system state
Δf – model uncertainty
$\boldsymbol{H}$ – matrix of second order partial derivatives
J – control quality criterion
J_{IAE} – integral absolute error
J_{ITAE} – integral time multiplied absolute error
n – penalty function exponent
q – weighting factor
Q – modified performance index
Q_{IAE} – integral absolute error with penalty function
Q_{ITAE} – integral time multiplied absolute error with penalty function
s – switching variable
t – time
t_0 – initial time
t_f – time when the switching line/plane stops moving
$t_{f\,opt}$ – optimal value of time t_f
u – control signal
u_{max} – maximum admissible value of the control signal
v_{max} – maximum admissible value of the system velocity
$\boldsymbol{x}$ – state vector
$\boldsymbol{x}_d$ – demand trajectory
x_1, x_2, x_3 – state variables
x_{10}, x_{20}, x_{30} – initial conditions of the system

δ – lower bound of function $b(\boldsymbol{x}, t)$
γ, η, μ – positive constants

|.| – absolute value
det(.) – matrix determinant
$.^T$ – matrix or vector transpose
sat(.) – saturation function
sgn(.) – sign function
[. , .] – closed interval
(. , .) – open interval

Chapter 1
Introduction

The main purpose of control engineering is to steer the regulated plant in such a way that it operates in a required manner. The desirable performance of the plant should be obtained despite the unpredictable influence of the environment on all parts of the control system, including the plant itself, and no matter if the system designer knows precisely all the parameters of the plant. Even though the parameters may change with time, load and external circumstances, still the system should preserve its nominal properties and ensure the required behaviour of the plant. In other words, the principal objective of control engineering is to design control (or regulation) systems which are robust with respect to external disturbances and modelling uncertainty. This objective may be very well achieved using the sliding mode technique (Utkin, 1992; Zinober, 1994; Utkin *et al.*, 1999; Gao, 1993; Utkin, 1993; Bartolini *et al.*, 1997a; Edwards and Spurgeon, 1998; Yu, 1998; Misawa and Utkin, 2000; Kaynak 2001; Sabanovic *et al.*, 2004; Edwards *et al.*, 2006; Bartoszewicz and Patton, 2007; Bartolini *et al.*, 2008; Bartoszewicz *et al.*, 2008; Shtessel *et al.*, 2008) which is the main subject of this monograph. To be more precise, in the monograph we focus our attention on only one – but very important aspect – of the sliding mode system design, i.e. we deal with the problem of the sliding plane selection. However, in order to make the text self-contained, we begin this chapter with presenting the main notions and concepts used in the field of variable structure systems and sliding mode control.

1.1 Preliminaries

In this section we present basic concepts and definitions which will be used further in the book. First, we introduce the concept of variable structure system (VSS), and then we present the notions of sliding mode and sliding mode control (SMC) which are crucial for the issues presented in the main part of this book, i.e. in chapters two and three. Finally, this section also presents the basic properties of variable structure systems with sliding modes which make these systems a feasible option for many control applications, especially those which require good robustness with respect to model uncertainty and external disturbances.

1.1.1 Variable Structure Systems

In recent years much of the research in the area of control theory focused on the design of discontinuous feedback which switches the structure of the system according to the evolution of its state vector. This control idea may be illustrated by the following example.

Let us consider the second order system

$$\dot{x}_1 = x_2$$
$$\dot{x}_2 = x_2 + u_i, \qquad i = 1, 2 \qquad (1.1)$$

where $x_1(t)$ and $x_2(t)$ denote the system state variables, with the following two feedback control laws

$$u_1 = f_1(x_1, x_2) = -x_2 - x_1, \qquad (1.2)$$

$$u_2 = f_2(x_1, x_2) = -x_2 - 4x_1. \qquad (1.3)$$

The performance of system (1.1) controlled according to (1.2) is shown in figure 1.1, and figure 1.2 presents the behaviour of the same system with feedback control (1.3). It can be clearly seen from those two figures that each of the feedback control laws (1.2) and (1.3) ensures the system stability only in the sense of Lyapunov.

Fig. 1.1 Phase portrait of system (1.1) with controller (1.2)

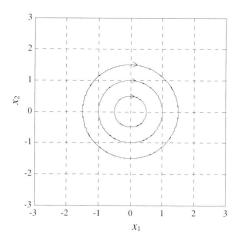

However, if the following switching strategy is applied

$$i = \begin{cases} 1 & \text{for} \quad \min\{x_1, x_2\} < 0 \\ 2 & \text{for} \quad \min\{x_1, x_2\} \geq 0 \end{cases}, \qquad (1.4)$$

then the system becomes asymptotically stable. This is illustrated in figure 1.3.

Moreover, it is worth to point out that system (1.1) with the same feedback control laws may exhibit completely different behaviour (and even become unstable). For example, if the switching strategy (1.4) is modified as

1.1 Preliminaries

Fig. 1.2 Phase portrait of system (1.1) with controller (1.3)

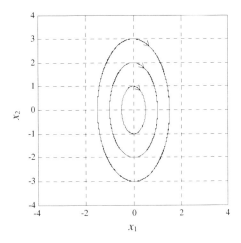

Fig. 1.3 Phase portrait of system (1.1) when switching strategy (1.4) is applied

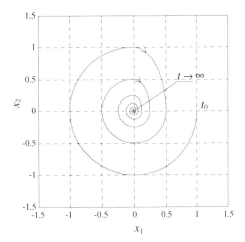

$$i = \begin{cases} 1 & \text{for} \quad \min\{x_1, x_2\} \geq 0 \\ 2 & \text{for} \quad \min\{x_1, x_2\} < 0 \end{cases}, \quad (1.5)$$

then the system output increases to infinity. The system dynamic behaviour, in this situation, is illustrated in figure 1.4.

This example presents the concept of variable structure control (VSC) and stresses that the system dynamics in VSC is determined not only by the applied feedback controllers but also, to a large extent, by the adopted switching strategy.

VSC is inherently a nonlinear technique and as such, it offers a variety of advantages which cannot be achieved using conventional linear controllers. Our next example shows one of those favourable features – namely it demonstrates that VSC may enable finite time error convergence. In this example, again we consider system (1.1) however now we apply the following controller

$$u = -x_2 - a\,\mathrm{sgn}(x_1) - b\,\mathrm{sgn}(x_2) \quad (1.6)$$

Fig. 1.4 Phase portrait of system (1.1) when switching strategy (1.5) is applied

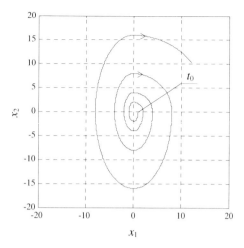

where $a > b > 0$. Closer analysis of the behaviour of system (1.1) with control law (1.6) demonstrates that, in this example, the system error converges to zero in finite time which can be expressed as

$$T = \frac{a}{b}\sqrt{2x_{10}}\left(\frac{1}{\sqrt{a-b}} + \frac{1}{\sqrt{a+b}}\right) \qquad (1.7)$$

where x_{10} and $x_{20} = 0$ represent initial conditions of system (1.1). Even though the error converges to zero in finite time, the number of oscillations in the system tends to infinity, with the period of the oscillations decreasing to zero. This is illustrated in figures 1.5 and 1.6. In the simulation example presented in the figures, the following parameters are used $a = 7$, $b = 3$, $x_{10} = 20$ and $x_{20} = 0$. Consequently, the system error is nullified at the time instant $T = 12.045$ and remains equal to zero for any time greater than T. Clearly these favourable properties are achieved using finite control signal. This controller, due to the way the phase trajectory – shown in figure 1.5 – is drawn, is usually called "twisting controller". It also serves as a good, simple example of the second order sliding mode controllers.

1.1.2 Sliding Mode Control

The two examples presented up to now demonstrate the principal properties of VSC systems. However, the main advantage of the systems is obtained when the controlled plant exhibits the sliding motion (Utkin, 1977; DeCarlo *et al.*, 1988; Slotine and Li, 1991; Hung *et al.*, 1993). The idea of sliding mode control (SMC) is to employ different feedback controllers acting on the opposite sides of a predetermined surface in the system state space. Each of those controllers pushes the system representative point (RP) towards the surface, so that the RP approaches the surface, and once it hits the surface for the first time it stays on it ever after. The resulting motion of the system is restricted to the surface, which graphically can be interpreted as "sliding" of the system RP along the surface. This idea is illustrated by the following example.

1.1 Preliminaries

Fig. 1.5 Phase portrait of system (1.1) controlled according to (1.6)

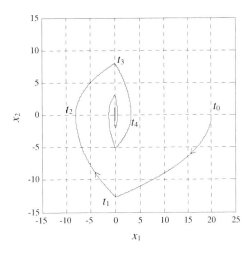

Fig. 1.6 State variables of system (1.1) controlled according to (1.6)

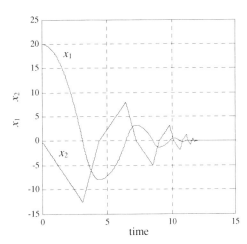

Let us consider another second order plant

$$\begin{aligned}\dot{x}_1 &= x_2 \\ \dot{x}_2 &= b\cos(mx_1)+u \qquad |b|<1\end{aligned} \qquad (1.8)$$

where b and m are possibly unknown constants. We select the following line in the state space

$$s = x_2 + cx_1 = 0 \qquad (1.9)$$

where c is a constant and apply the controller

$$u = -cx_2 - \operatorname{sgn}(s). \qquad (1.10)$$

In this equation sgn(.) function represents the sign of its argument, i.e. $sgn(s < 0) = -1$ and $sgn(s > 0) = +1$. With this controller the system representative point moves towards line (1.9) always when it does not belong to the line. Then, once it hits the line, the controller switches the plant input (in the ideal case) with infinite frequency. Therefore, line (1.9) is called the switching line. Furthermore, since after reaching the line, the system RP slides along it, then the line is also called the sliding line. This example is illustrated in figure 1.7.

The system parameters used in the presented simulation are $c = 0.5$, $b = 0.75$, $m = 10$ and the simulation is performed for the following initial conditions $x_{10} = 5$ and $x_{20} = 1$. Notice that when the plant remains in the sliding mode, its dynamics is completely determined by the switching line (or in general the switching hypersurface) parameters. This implies that neither model uncertainty nor matched external disturbance affects the plant dynamics (Draženović, 1969) which is a highly desirable system property. This property can also be justified geometrically, if one notices that in our example the slope of line (1.9) fully governs the plant motion in the sliding mode. Therefore, in SMC systems we usually make the distinction between two phases: the first one – called the reaching phase – lasts until the controlled plant RP hits the switching surface, and the second one – the sliding phase – begins when the RP reaches the surface. In the latter phase the plant insensitivity to a class of modelling inaccuracies and external disturbances is ensured. Let us stress that the system robustness with respect to unmodelled dynamics, parameter uncertainty and external disturbances is guaranteed only in the sliding mode. Therefore, shortening or (if possible) even complete elimination of the reaching phase is an important and timely research subject. Thus, in the next section we briefly present some recent results concerning this issue.

Another immediate consequence of the fact that in the sliding mode, the system RP is restricted to the switching hypersurface (which is a subset of the state space) is reduction of the system order. If the system of the order n has m independent inputs, then the sliding mode takes place on the intersection of m hypersurfaces and the reduced order of the system is equal to the difference $n - m$. To be more precise, in multi-input systems the sliding mode may take place either independently on each switching hypersurface or only on the intersection of the surfaces. In the first case the system RP approaches each surface at any time instant and once it hits any of the surfaces it stays on this surface ever after. This scenario is shown in figure 1.8. In the latter case, however, the system RP does not necessarily approach each of the surfaces, but it always moves towards their intersection. In this case, which is illustrated in figure 1.9, the system RP may hit a surface and move away from it (might possibly cross a switching surface), but once it reaches the intersection of all the surfaces, then the RP never leaves it.

As it has already been mentioned, the switching surface completely determines the plant dynamics in the sliding mode. Therefore, selecting this surface (Gao, *et al.*, 1995; Ignaciuk and Bartoszewicz, 2008a; Ignaciuk and Bartoszewicz, 2008b; Janardhanan and Kariwala, 2008) is one of the two major tasks in the process of the SMC system design. In order to stress this issue let us point out that the same controller which has been considered in the last example may result in a very different system performance, if the sliding line slope c is selected in another way.

1.1 Preliminaries

Fig. 1.7 Phase trajectory of system (1.1) controlled according to (1.10)

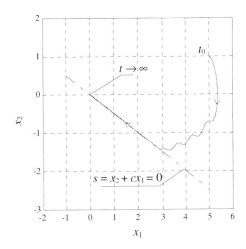

This can be easily noticed if one takes into account any negative c. Then, controller (1.10) still ensures stability of the sliding motion on line (1.9), i.e. the system RP still converges to the line, however the system is unstable since both state variables x_1 and x_2 tend to (either plus or minus) infinity while the system RP slides away from the origin of the phase plane along line (1.9).

The other major task in the SMC system design is the selection of an appropriate control law. This can be achieved either by assuming a certain kind of the control law (usually motivated by some previous engineering experience) and proving that this control satisfies one of the so-called reaching conditions or by applying the reaching law approach. The reaching conditions (Edwards and Spurgeon, 1998) ensure stability of the sliding motion and therefore they are naturally derived using Lyapunov stability theory. On the other hand, if the reaching law approach is adopted for the purpose of a sliding mode controller construction (Hung et al., 1993), then a totally different design philosophy is employed. In this case the desired evolution of the switching variable s is specified first, and then a control law ensuring that s changes according to the specification is determined.

Sliding mode controllers guarantee system insensitivity with respect to matched disturbance and model uncertainty, and cause reduction of the plant order. Moreover, they are computationally efficient, and may be applied to a wide range of various, possibly nonlinear and time-varying plants. However, often they also exhibit a serious drawback which essentially hinders their practical applications. This drawback – high frequency oscillations which inevitably appear in any real system whose input is supposed to switch infinitely fast – is usually called chattering. If system (1.8) exhibits any, even arbitrarily small, delay in the input channel, then control strategy (1.10), will cause oscillations whose frequency and amplitude depend on the delay. With the decreasing of the delay time, the frequency rises and the amplitude gets smaller. This is a highly undesirable phenomenon, because it causes serious wear and tear on the actuator components. Therefore, a few

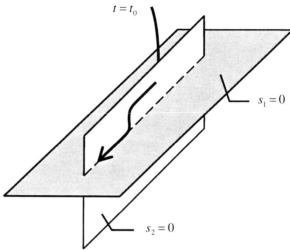

Fig. 1.8 Independent sliding motion on each switching surface

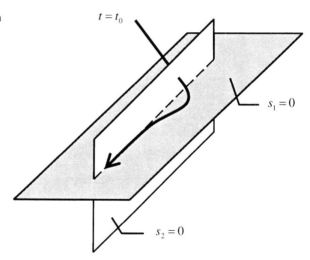

Fig. 1.9 Sliding motion on the intersection of the switching surfaces

methods to eliminate chattering have been proposed. The most popular of them uses function

$$\operatorname{sat}(s) = \begin{cases} -1 & \text{for } s < -\rho \\ \dfrac{1}{\rho} s & \text{for } |s| \leq \rho \\ 1 & \text{for } s > \rho \end{cases} \quad (1.11)$$

(where ρ is a positive, usually small constant) instead of sgn(s) in the definition of the discontinuous control term. With this modification the term becomes continuous and the switching variable does not converge to zero but to the closed interval

1.1 Preliminaries

Fig. 1.10 Function sat(s)

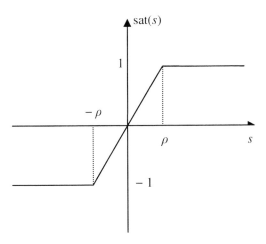

$[-\rho, \rho]$. Consequently, the system RP after the reaching phase termination, belongs to a layer around the switching hyperplane and therefore this strategy is called boundary layer controller (Slotine and Li, 1991).

Other approaches to the chattering elimination include:

a) introduction of other nonlinear approximations of the discontinuous control term, for example the so called fractional approximation defined as

$$\text{approx}(s) = \frac{s}{\rho + |s|} \qquad (1.12)$$

(Ambrosino et al., 1984; Burton and Zinober, 1986; Spurgeon and Davies, 1993; Yu and Lloyd, 1997);
b) replacing the boundary layer with a sliding sector (Shyu et al., 1992; Xu et al., 1996);
c) using dynamic sliding mode controllers (Sira-Ramirez, 1993a; Sira-Ramirez, 1993b; Sira-Ramirez and Llanes-Santiago, 1994; Zlateva, 1996; Bartolini and Pydynowski, 1996; Rios-Bolivar et al., 1997; Spurgeon and Lu, 1997; Lu and Spurgeon, 1998; Selisteanu et al., 2007);
d) using fuzzy sliding mode controllers (Palm, 1994; Palm et al. 1997; Choi and Kim, 1997);
e) using second (or higher) order sliding mode controllers (Levant, 1993; Bartolini et al., 1997b; Bartolini et al., 1997c; Bartolini et al., 1998; Levant, 2003).

The phenomenon of chattering has been extensively analysed in many papers using describing function method and various stability criteria (Shtessel and Lee, 1996; Chung and Lin, 1999; Bartoszewicz, 2000).

In this section we presented a concise introduction to the sliding mode control of continuous time systems. We did not attempt to make our presentation particularly in-depth, exhaustive or complete, but we tried to give a brief background for the issues presented further in this monograph. Indeed in this section we explained some basic notions which we will use in chapters 2 and 3 for the design of the moving switching surfaces for sliding mode control of continuous time systems.

1.2 Previous Work on Sliding Mode Control with Time-Varying Switching Surfaces

In this section we present a brief overview of the most important, recent results in the field of sliding mode control with time-varying switching surfaces. These surfaces help eliminate the reaching phase from variable structure control systems and ensure plant insensitivity with respect to external disturbance and model uncertainty from the very beginning of the control process.

In fact, time-varying sliding surfaces have recently been applied both in continuous (Chang and Hurmuzlu, 1992; Chang and Hurmuzlu, 1993; Choi et al., 1993; Choi and Park, 1994; Choi et al., 1994; Bartoszewicz, 1995; Bartoszewicz, 1996; Ghosh and Olgac, 1997; Kim et al., 1998; Temeltas, 1998; Weisheng et al., 1998; Ha et al., 1999; Park and Choi, 1999; Park et al., 1999; Demin et al., 2000; Yilmaz and Hurmuzlu, 2000; Betin et al., 2002a; Betin et al., 2002b; Chen et al., 2002; Iliev and Hristozov, 2002; Iliev and Kalaykov, 2002; Tokat et al., 2002; Tokat et al., 2003a; Tokat et al., 2003b; Iliev and Kalaykov, 2004; Sivert et al., 2004a; Sivert et al., 2004b; Sivert et al., 2004c; Jinggang et al., 2006; Corradini and Orlando, 2007) and discrete time systems (Bartoszewicz, 1998; Zhang and Guo, 2000; Chakravarthini Saaj and Bandyopadhyay, 2001; Chung et al., 2004; Jinggang et al., 2004a; Jinggang et al., 2004b). Most of these papers are devoted to the second order plant control. Therefore, the problem of the sliding hypersurface design – in those papers – is reduced to the issue of designing appropriately selected time-varying sliding lines. One of the first applications of the time-varying switching lines for the second order plant control is reported in (Choi et al., 1993, Choi and Park, 1994; Choi et al., 1994). The authors of those papers propose using switching lines which instantaneously change their position on the phase plane. The lines can move in one of the two ways: they may either rotate around the origin of the plane, or they may be shifted without changing their slope. This approach results in faster error convergence, than the one obtained in conventional systems with fixed, time-invariant sliding lines. However, as it is demonstrated in (Bartoszewicz, 1995), application of these lines does not ensure complete insensitivity of the plant with respect to external disturbance and modelling uncertainty. Therefore, the switching lines which smoothly move on the phase plane from their original location to the final, predefined position are introduced in papers (Bartoszewicz, 1995; Bartoszewicz, 1996). Furthermore, in paper (Bartoszewicz, 1996) application of the time-varying switching curves, which results in finite time error convergence is also proposed. Another approach similar to that presented in (Bartoszewicz, 1995; Bartoszewicz, 1996) is also investigated in several other papers (Park et al., 1999; Betin et al., 2002a; Betin et al., 2002b; Tokat et al., 2002; Tokat et al., 2003b; Sivert et al., 2004a; Sivert et al., 2004b; Sivert et al., 2004c; Bartoszewicz and Żuk, 2006). In (Betin et al., 2002a; Betin et al., 2002b) the sliding mode control with time-varying switching lines is applied to the position control of a dc motor. The analysis presented in those two papers is illustrated with experimental results. Laboratory tests are also reported in (Sivert et al., 2004a; Sivert et al., 2004b), where the time-varying switching lines are applied to the position control of induction machines. For that purpose, the switching line

1.2 Previous Work on Sliding Mode Control

which is shifted with a constant velocity to its predetermined final position is applied. The parameters describing the line are selected so that the current constraint is satisfied and the integral absolute shaft position error is minimised. Also in (Sivert et al., 2004c) time-varying switching lines are applied to the shaft position control of electric motors. First, the moving switching lines are applied for the dc motor regulation, and then for the induction motor control. The theoretical results obtained in the paper are verified in a series of experiments which enable the comparison of the sliding mode control with time-varying and with conventional, fixed switching lines. The results of these experiments again confirm the advantages of sliding mode control with moving switching lines, i.e. its insensitivity with respect to external disturbance and modelling uncertainty from the very beginning of the control process. Then in paper (Park et al., 1999), time-varying (either rotating or shifted) switching lines are applied to control the position of an n-degree of freedom robot manipulator. The authors of that paper propose to substitute the discontinuous part of the sliding mode controller with a PID type structure, which enables to eliminate the undesirable chattering phenomenon. Moreover, this solution helps nullify the steady state system error in the situation when the controlled plant is subject to a constant disturbance. The performance of the controller proposed in that paper is illustrated with an example of a two degree of freedom, stiff manipulator position regulation. Also in paper (Bartoszewicz and Żuk, 2006) an application of the sliding mode controller with a moving switching line for the two degree of freedom robot arm position regulation is proposed. In that paper, velocity and acceleration constraints are explicitly taken into account. Further results of the application of time-varying switching lines to the sliding mode control of the continuous time, second order dynamic systems are presented in (Tokat et al., 2002; Tokat et al., 2003b). In a paper (Tokat et al., 2002), its authors introduce a new coordinate frame, and then they use the frame to define a rotating straight switching line. The same authors in paper (Tokat et al., 2003b) propose another type of time-varying switching line, i.e. a changing shape parabola.

A new way of moving the switching hypersurfaces is presented in (Chang and Hurmuzlu, 1992; Chang and Hurmuzlu, 1993) and (Yilmaz and Hurmuzlu, 2000). The authors of those papers show that applying hyperplanes which exponentially approach some predefined, fixed position helps eliminate the reaching phase. Their solution ensures optimal, in the considered class of the switching hyperplanes, transient performance of the controlled system. In (Chang and Hurmuzlu, 1992; Chang and Hurmuzlu, 1993) the authors apply the exponentially moving switching lines to control a two link robot manipulator, and also to regulate the position of a bipedal, five link walking robot. The widest spectrum of the results, among contributions (Chang and Hurmuzlu, 1992; Chang and Hurmuzlu, 1993; Yilmaz and Hurmuzlu, 2000), is presented in paper (Yilmaz and Hurmuzlu, 2000). In that paper the results of (Chang and Hurmuzlu, 1992; Chang and Hurmuzlu, 1993) are generalised, so that they apply not only to the second order plants but also to the plants of arbitrary order, possibly higher than two. Moreover, in the same paper (Yilmaz and Hurmuzlu, 2000) an optimisation criterion, taking into account not only the regulation error and its derivatives but also the magnitude of the control signal is introduced and the parameters of the switching hyperplanes are selected in such a way that the criterion is minimised.

The idea of using the time-varying switching hypersurfaces in the sliding mode control of continuous time n-th order dynamic system is also indirectly exploited in (Ma et al., 1999). The authors of that work introduce the moving switching hyperplanes and propose appropriate shaping of the demand trajectory. Furthermore, they precisely analyse an example of a single link robot manipulator and they demonstrate that their regulation strategy eliminates the reaching phase and consequently ensures better robustness of the manipulator with respect to disturbance and modelling uncertainty than the popular computed torque method (Spong and Vidyasagar, 1989). Also in papers (Weisheng et al., 1998; Demin et al., 2000) the control of the continuous n-th order system is considered. For that purpose moving switching hypersurfaces ensuring almost time optimal performance of the controlled systems are applied. That approach results in faster error convergence than the convergence obtained in conventional VSC and makes the system insensitive with respect to disturbance and modelling uncertainty, from the very beginning of the control process. In a paper (Demin et al., 2000) the considered method is applied to the control of an autonomous underwater vehicle. The paper demonstrates that the method not only ensures system robustness but also helps reduce the undesirable chattering phenomenon. Moreover, in paper (Weisheng et al., 1998) the same method is applied to the control of an arbitrary, time-varying nonlinear plant. In both papers (Demin et al., 2000) and (Weisheng et al., 1998) the control signal constraint is explicitly taken into account. The solutions proposed in that papers result in the time optimal control for the second order systems and quasi time optimal control for higher order plants. Moreover, paper (Ghosh and Olgac, 1997) presents further analysis of the results given in three manuscripts (Choi et al., 1993, Choi and Park, 1994; Choi et al., 1994) already briefly referred to at the beginning of this survey. Paper (Ghosh and Olgac, 1997) contains a generalisation of the results presented before by (Choi et al., 1993; Choi and Park, 1994; Choi et al., 1994) for the plants of arbitrary order n. It demonstrates how the mechanism of the instantaneous sliding hypersurface position change can be defined in the n-dimensional phase space.

All the papers which we have already mentioned (Chang and Hurmuzlu, 1992; Chang and Hurmuzlu, 1993; Choi et al., 1993; Choi et al., 1994; Choi and Park, 1994; Bartoszewicz, 1995; Bartoszewicz, 1996; Ghosh and Olgac, 1997; Weisheng, et al., 1998; Ma et al., 1999; Park et al., 1999; Demin et al., 2000; Yilmaz and Hurmuzlu, 2000; Betin et al., 2002a; Betin et al., 2002b; Tokat et al., 2002; Sivert et al., 2004a; Sivert et al., 2004b; Sivert et al., 2004c; Tokat et al., 2003b) are concerned with the application of time-varying switching hypersurfaces, and predominantly switching lines, for the sliding mode control of continuous time dynamic systems. However, the moving hypersurfaces may also be applied in discrete time systems (Bartoszewicz, 1998; Zhang and Guo, 2000; Chakravarthini Saaj and Bandyopadhyay, 2001; Chung et al., 2004; Jinggang et al., 2004a; Jinggang et al., 2004b). For example, Zhang and Guo in paper (Zhang and Guo, 2000) introduced rotating switching lines into the discrete time sliding mode control system of a hard disk drive. The system eliminates chattering, and ensures quasi time optimal transient performance, together with the controlled plant insensitivity to

1.2 Previous Work on Sliding Mode Control

external disturbance and the possible plant parameter changes. These properties are verified in simulation examples. Time optimal control of a hard disk drive is also considered in paper (Chung *et al.*, 2004). The authors of that paper propose to divide the control process into three phases, introduce a different control algorithm in each of them, and apply a time-varying sliding line in one of the three phases. Their control may be applied in time-varying parameter control systems. Then in paper (Jinggang *et al.*, 2004b), the sliding mode control using the rotating switching line is proposed for the general second order discrete time system. The line rotates from its original position imposed by the initial conditions of the system, to the final predetermined location on the phase plane. The proposed control scheme is applied for the dc motor shaft position regulation. The results reported in that paper confirm system insensitivity with respect to the changing parameters of the controlled plant and external disturbance, right from the beginning of the regulation process. Furthermore, in paper (Jinggang *et al.*, 2004a) a control system with two switching lines is considered. One of these lines is shifted on the phase plane and the other one remains time-invariant. At the initial time, the representative point describing the system dynamics belongs to the moving line, and later on when it arrives close enough to the time-invariant line, it starts moving towards the origin of the state space along the latter line. This approach, as demonstrated in the presented simulation tests, ensures insensitivity of the system during the whole regulation process. We conclude this brief reference overview with just a mention of paper (Chakravarthini Saaj and Bandyopadhyay, 2001) which applies the reaching law approach (Gao *et al.*, 1995) together with the rotating switching line for the control of the discrete time second order plant. The main advantage of this approach is that it does not require both state variables to be available for measurement, but uses only the output feedback. The state variable measurement is eliminated by an appropriate increase of the sampling frequency in the system (i.e. multirate sampling). The state variables are required neither to calculate any of the feedback signals generated in the system, nor to find out the switching function value determining which of the feedback signals is used at a particular moment. The paper demonstrates, both analytically and by means of computer simulations, that the proposed method helps shorten the reaching phase.

The authors of all the papers already presented in this section apply the moving switching surfaces in order to shorten or eliminate the reaching phase and by this means to improve the controlled plant robustness at the initial stage of the regulation process. However, they usually do not take into account the physical constraints which are inevitable in any real system. In other words, the authors of the majority of the papers assume that the control signal may take arbitrarily big values and that the system state variables are not subject to any constraints. Since in control engineering practice this is almost never the case, in this book we explicitly take into account state and input signal constraints and we design the optimal, in the sense of two standard control quality criteria, switching planes for the sliding mode control of the second and the third order nonlinear and possibly time-varying plants. The criteria considered in this book are integral absolute error (IAE) and integral time multiplied absolute error (ITAE). We focus our attention on the second and the third order plants since, on one hand, they are still mathematically tractable and on the other, they are fairly typical in the control engineering domain. Indeed most of electric, pneumatic or hydraulic drives and actuators, electric power converters, automotive active suspension systems, robot manipulators, etc. belong to this class.

The main essence of this monograph is presented in chapters 2 and 3. These two chapters comprise only *original results* obtained recently by the authors. Some of the results have already been published in several journal (Bartoszewicz, 1995; Bartoszewicz, 1996; Bartoszewicz and Nowacka, 2004a; Bartoszewicz and Nowacka, 2005b; Bartoszewicz and Nowacka, 2005c; Bartoszewicz and Nowacka, 2006b; Bartoszewicz and Nowacka, 2006c; Bartoszewicz and Nowacka, 2006d; Bartoszewicz and Żuk, 2006; Bartoszewicz and Nowacka, 2007a; Bartoszewicz and Nowacka-Leverton, 2007) and conference papers (Bartoszewicz and Nowacka, 2004b; Bartoszewicz and Nowacka, 2005a; Nowacka and Bartoszewicz, 2005; Bartoszewicz and Nowacka, 2006a; Nowacka and Bartoszewicz, 2006a; Nowacka and Bartoszewicz, 2006b; Nowacka and Bartoszewicz, 2006c; Nowacka and Bartoszewicz, 2006d; Bartoszewicz and Nowacka, 2007b; Nowacka-Leverton and Bartoszewicz, 2008a; Nowacka-Leverton and Bartoszewicz, 2008b; Nowacka-Leverton and Bartoszewicz, 2008c), while various other derivations and conclusions presented in the two chapters have not been available in print yet.

The main part of the book is organised as follows. Chapter 2 is devoted to the optimal switching line design for the second order plant. The chapter begins with the precisely formulated plant description and the problem statement. Then, in section 2.1 we introduce the idea of the moving switching lines and we present the control strategy considered further in that chapter. In particular, in section 2.1 we derive the equations describing the system tracking error evolution and its derivative. Moreover, in the same section we demonstrate that the considered plant controlled according to the proposed strategy exhibits no overshoots or oscillations and is insensitive with respect to external disturbance and modelling uncertainty from the very beginning of the regulation process. Next in section 2.2 we move on to the design of the optimal switching line. At the beginning of that section we derive the formula which gives the integral absolute error and then we minimise this formula taking into account different constraints. In section 2.2.1 we consider only the input signal constraint, in section 2.2.2 only the velocity (or more precisely the first derivative of the output signal) constraint and then in section 2.2.3 we require that both the input signal and the velocity constraints hold at the same time. Each of the three cases is illustrated by a separate simulation example. Then in section 2.3 we again design the optimal switching line for the same second order plant, but this time we take into account another control quality criterion – the integral time multiplied absolute error. First we calculate this criterion, and then in sections 2.3.1, 2.3.2 and 2.3.3 we minimise the criterion taking into account input signal constraint only, then velocity (the first derivative of the output signal) constraint only, and finally both of these constraints simultaneously. Similarly as in the first part of chapter 2, also the results derived in sections 2.3.1, 2.3.2 and 2.3.3 are extensively verified by appropriate simulation examples.

1.2 Previous Work on Sliding Mode Control

Chapter 3 is entirely concerned with the control of the third order system. In that chapter we select the switching planes which ensure optimal, in the sense of integral absolute error and integral time multiplied absolute error, performance of the system subject to three constraints. The constraints considered in that chapter are imposed on: the input signal, the first derivative of the plant output and the second derivative of the system output. The last two signals are – similarly as in chapter 2 – referred to as the system velocity and acceleration, respectively. This is justified by the fact that in mechanical systems they are closely related to the output signal rate of change (velocity) and its acceleration. This is particularly true when the demand output signal of the system is constant, for example in the so called point to point (PTP) control of robot manipulator position.

At the beginning of chapter 3 we define the considered plant and the control objectives. Then in section 3.1 we introduce the proposed time-varying switching plane and we specify the control signal which maintains the system representative point on the plane. Further in the same section we derive the equations describing the tracking error of the system and its first and second derivatives. Next, in section 3.2 we calculate the integral absolute error in the considered system and we minimise it subject to the input signal, acceleration and velocity constraints. As a result of the minimisation procedure we obtain the optimal (in the sense of the integral absolute error) parameters of the considered switching plane. This minimisation procedure is performed in sections 3.2.1, 3.2.2 and 3.2.3 for each of the constraints separately, then in sections 3.2.4, 3.2.5 and 3.2.6 for each pair of the constraints at a time (in section 3.2.4 acceleration and velocity constraints are taken into account, in section 3.2.5 acceleration and input signal constraints are considered and in section 3.2.6 input signal and velocity constraints are analysed), and finally in section 3.2.7 for all the three constraints simultaneously. Just as it is done in chapter 2, also all the derivations presented in section 3.2, are illustrated with carefully selected simulation examples.

The second part of chapter 3, i.e. section 3.3 is organised similarly to section 3.2. Again it is divided into seven subsections each of them devoted to the minimisation of the same criterion, but subject to different constraints. The criterion considered in this section is the integral time multiplied absolute error. Sections 3.3.1, 3.3.2 and 3.3.3 present the solution of the minimisation problem with each of the three constraints considered one by one, later on sections 3.3.4, 3.3.5 and 3.3.6 show the same solution but with each two of the constraints taken at a time, and at last section 3.2.7 demonstrates the solution of this minimisation task when all the three constraints are dealt with all together. The results obtained in section 3.3, i.e. the time-varying switching planes optimal in the sense of integral time multiplied absolute error are quite similar, and sometimes even exactly the same, as the results found earlier in section 3.2 where the switching planes optimal in the sense of integral absolute error are designed. Therefore, we confine our presentation in section 3.3 to the derivations, comments and extensive analysis, and we do not repeat unchanged or only very slightly modified simulation examples already introduced in section 3.2.

Both in chapters 2 and 3 we consider not only the conventional constraints expressed by inequalities, but we also take into account more convenient, elastic (or stretchable) constraints which in industrial applications often turn out to be more practical than the conventional ones. These constraints tolerate (or allow for) some, usually minor increase of the particular signal under consideration (input signal, velocity or acceleration) beyond its threshold value, if this increase does not jeopardise safety of operation of the controlled plant but results in a major improvement of its dynamic performance. In other words, the elastic constraints help trade off system dynamic performance and reasonably increased values of the input signal, velocity or acceleration. This is a fairly feasible approach since in engineering applications, constraints are quite rarely given as crisp values which under no circumstances can be exceeded. More often they are specified as indicators with some accuracy bounds to be taken into account.

Finally, chapter 4 presents the conclusions of the book. In that chapter we establish some links and relations between the results presented in this monograph and other ideas already well established in the field of sliding mode control. We also make an attempt to envision future trends and promising research directions in the field directly related to the subject of this book.

Chapter 2
Time-Varying Sliding Modes for the Second Order Systems

In this chapter we consider the time-varying and nonlinear, second order system described by the following equations

$$\dot{x}_1 = x_2$$
$$\dot{x}_2 = f(x,t) + \Delta f(x,t) + b(x,t)u + d(t)$$

(2.1)

where x_1, x_2 are the state variables of the system and

$$x(t) = [x_1(t)\; x_2(t)]^T$$

(2.2)

is the state vector, t denotes time, u is the input signal, b, f – are a priori known, bounded functions of time and the system state, Δf and d are functions representing the system uncertainty and external disturbances, respectively. Further in this book, it is assumed that there exists a strictly positive constant δ which is the lower bound of the absolute value of $b(x, t)$, i.e.

$$0 < \delta = \inf\{|b(x, t)|\}.$$

(2.3)

Furthermore, functions Δf and d are unknown and bounded. Therefore, there exists a constant μ which for every pair (x, t) satisfies the following inequality

$$|\Delta f(x, t) + d(t)| \leq \mu.$$

(2.4)

The initial conditions of the system are denoted as x_{10}, x_{20} where

$$x_{10} = x_1(t_0),\; x_{20} = x_2(t_0).$$

(2.5)

System (2.1) is supposed to track the desired trajectory given as a function of time

$$x_d(t) = [x_{1d}(t)\quad x_{2d}(t)]^T$$

(2.6)

where $x_{2d}(t) = \dot{x}_{1d}(t)$ and $x_{2d}(t)$ is a differentiable function of time. The trajectory tracking error is defined by the following vector

$$e(t) = [e_1(t)\; e_2(t)]^T = x(t) - x_d(t).$$

(2.7)

Hence, we have

$$e_1(t) = x_1(t) - x_{1d}(t)$$

(2.8)

and

$$e_2(t) = x_2(t) - x_{2d}(t).$$

(2.9)

A. Bartoszewicz and A. Nowacka-Leverton: Time-Varying Sliding Modes, LNCIS 382, pp. 17–65.
springerlink.com   © Springer-Verlag Berlin Heidelberg 2009

In this chapter it is assumed that at the initial time $t = t_0$, the tracking error and the error derivative can be expressed as

$$e_1(t_0) = e_0 \neq 0, e_2(t_0) = 0 \qquad (2.10)$$

where e_0 is an arbitrary real number different from zero. This condition is indeed satisfied in many practical applications such as position control or set point change of second order systems. An example of these applications is point to point (PTP) control of robot manipulators, that is moving the manipulator arm from its initial location where it is originally at a halt, to another predefined position at which the arm stops and again is expected to remain at rest.

Further in this chapter, we present a detailed description of the sliding mode control strategy which ensures optimal performance of the system and its robustness with respect to both the system uncertainty $\Delta f(\mathbf{x}, t)$ and external disturbance $d(t)$.

2.1 Control Strategy

In order to effectively control system (2.1), i.e. to eliminate the reaching phase and to obtain system insensitivity with respect to both external disturbance $d(t)$ and the model uncertainty $\Delta f(\mathbf{x}, t)$ from the very beginning of the system motion, we introduce a time-varying switching line. The line slope does not change during the control process, which implies that the line moves on the phase plane without rotating. In other words, the line is shifted in the state space with a constant angle of inclination. At the beginning the line moves uniformly (i.e. with a constant velocity) in the state space and then it stops at a time instant $t_f > t_0$. Consequently, for any $t \in \langle t_0, t_f \rangle$ the switching line is described by the following equation

$$s(\mathbf{e}, t) = 0 \quad \text{where} \quad s(\mathbf{e}, t) = e_2(t) + ce_1(t) + A + Bt \qquad (2.11)$$

where c, A and B are constants. The selection of these constants will be considered further in this chapter. Since the considered line stops moving at the time t_f, for any $t \geq t_f$ it is fixed and can be described as follows

$$s(\mathbf{e}, t) = 0 \quad \text{where} \quad s(\mathbf{e}, t) = e_2(t) + ce_1(t). \qquad (2.12)$$

As it has been pointed out in the previous chapter, in order to ensure system (2.1) stability in the sliding motion on the line described by equations (2.11) and (2.12), parameter c in these equations must be strictly positive, i.e.

$$c > 0. \qquad (2.13)$$

The switching line considered in this chapter is illustrated in figure 2.1. Assuming that $t_0 = 0$, the initial horizontal, i.e. measured in the direction of the e_1 axis, offset of the line, equals $-A/c$ and the line moves along this axis at a rate $-B/c$. The same motion can also be interpreted as a shift along the vertical (e_2) axis at a speed of $-B$ with initial vertical offset equal to $-A$. Furthermore, the distance between the switching line and the origin of the state space measured along the shortest path, initially equal to $|A|/\sqrt{1+c^2}$, decreases at a velocity

2.1 Control Strategy

Fig. 2.1 Time-varying switching line

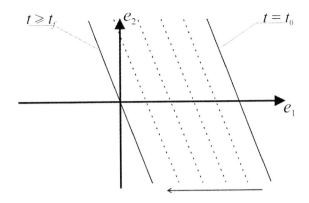

of $-|B|/\sqrt{1+c^2}$. Once the line reaches the origin of (e_1, e_2) coordinate frame, the line stops moving and from then on it remains fixed and time-invariant.

Furthermore, in order to actually eliminate the reaching phase, and consequently to ensure insensitivity of the considered system from the very beginning of its motion, the constants A, B and c should be chosen in such a way that the representative point of the system at the initial time $t = t_0$ belongs to the switching line. For that purpose, the following condition must be satisfied

$$s[e(t_0), t_0] = e_2(t_0) + ce_1(t_0) + A + Bt_0 = 0. \quad (2.14)$$

Notice that the input signal

$$u = \frac{-f(x,t) - ce_2(t) + \dot{x}_{2d}(t) - B - \gamma \operatorname{sgn}[s(e,t)]}{b(x,t)} \quad (2.15)$$

where $\gamma = \eta + \mu$ and η is a strictly positive constant, ensures the stability of the sliding motion on the switching line (2.11). In order to verify this property we consider the product

$$s(e,t)\dot{s}(e,t) = s(e,t)[\dot{e}_2(t) + ce_2(t) + B]. \quad (2.16)$$

Substituting relations (2.1), (2.7) and (2.15) into (2.16), we get

$$\begin{aligned}
s(e,t)\dot{s}(e,t) &= \\
&= s(e,t)\Big[f(x,t) + \Delta f(x,t) + b(x,t)u + d(t) - \dot{x}_{2d}(t) + ce_2(t) + B\Big] = \\
&= s(e,t)\Big\{f(x,t) + \Delta f(x,t) + b(x,t) \cdot \\
&\quad \cdot \{-f(x,t) - ce_2(t) + \dot{x}_{2d}(t) - B - \gamma \operatorname{sgn}[s(e,t)]\}/b(x,t) + \\
&\quad + d(t) - \dot{x}_{2d}(t) + ce_2(t) + B\Big\} = \\
&= s(e,t)\{\Delta f(x,t) + d(t) - \gamma \operatorname{sgn}[s(e,t)]\} \leq -\eta |s(e,t)|
\end{aligned} \quad (2.17)$$

which proves the existence and stability of the sliding motion on the line described by equation (2.11). Furthermore, in a very similar way, it can be verified that the input signal

$$u = \frac{-f(x,t) - ce_2(t) + \dot{x}_{2d}(t) - \gamma \text{sgn}\left[s(e,t)\right]}{b(x,t)} \tag{2.18}$$

ensures stable sliding motion of the considered system on the line determined by equation (2.12). Consequently, for any time $t \in \langle 0, t_f \rangle$ the system dynamics is described by equation (2.11) with initial conditions (2.10) and for $t \geq t_f$ it is governed by equation (2.12). Therefore, in order to find the system tracking error, now we consider the following equation

$$e_2(t) + ce_1(t) + A + Bt = 0 \tag{2.19}$$

which determines the considered switching line for any time $t \leq t_f$. In order to solve this equation, we take into account its homogenous counterpart

$$e_2(t) + ce_1(t) = 0 \tag{2.20}$$

which shows that the exponential term in the solution of equation (2.19) decays with the time constant equal to the inverse of c. Solving equation (2.19) with initial condition (2.10) and assuming for the sake of clarity that $t_0 = 0$ we get the tracking error and its derivative for the time $t \in \langle 0, t_f \rangle$

$$e_1(t) = \left(e_0 + \frac{A}{c} - \frac{B}{c^2}\right)e^{-ct} - \frac{A}{c} + \frac{B}{c^2} - \frac{B}{c}t, \tag{2.21}$$

$$e_2(t) = -c\left(e_0 + \frac{A}{c} - \frac{B}{c^2}\right)e^{-ct} - \frac{B}{c}. \tag{2.22}$$

Taking into account condition (2.14) and the assumption that $t_0 = 0$ we obtain

$$A = -ce_0. \tag{2.23}$$

Then formulae (2.21) and (2.22) can be written as

$$e_1(t) = -\frac{B}{c^2}e^{-ct} + e_0 + \frac{B}{c^2} - \frac{B}{c}t, \tag{2.24}$$

$$e_2(t) = \frac{B}{c}e^{-ct} - \frac{B}{c} = \frac{B}{c}\left(e^{-ct} - 1\right). \tag{2.25}$$

Notice that for the time $t \geq t_f$ the switching line is fixed and passes through the origin of the error state space. This leads to condition

$$A + Bt_f = 0. \tag{2.26}$$

2.1 Control Strategy

From equations (2.26) and (2.23) we obtain

$$t_f = \frac{e_0 c}{B}.$$

(2.27)

Inequality (2.13) and the fact that t_f is greater than zero imply that constants e_0 and B always have the same signs, i.e.

$$\operatorname{sgn}(B) = \operatorname{sgn}(e_0).$$

(2.28)

This is an important property which helps analyse the system behaviour. We will often use this property further in the text.

Next, we will analyse the behaviour of the system in the second phase of its motion, that is when the switching line does not move. The time invariant switching line is described by relation (2.12), which is equivalent to equation (2.20). The initial conditions which are necessary to solve equation (2.20) can be determined from equations (2.24) and (2.25) whose values are evaluated at the time instant t_f

$$e_1(t_f) = -\frac{B}{c^2} e^{-ct_f} + e_0 + \frac{B}{c^2} - \frac{B}{c} t_f = \frac{B}{c^2}\left(1 - e^{-ct_f}\right) + e_0 - \frac{B}{c} t_f,$$

(2.29)

$$e_2(t_f) = \frac{B}{c}\left(e^{-ct_f} - 1\right).$$

(2.30)

Then substituting equation (2.27) into formulae (2.29) and (2.30), we obtain

$$e_1(t_f) = \frac{B}{c^2}\left(1 - e^{-c^2 e_0/B}\right),$$

(2.31)

$$e_2(t_f) = \frac{B}{c}\left(e^{-c^2 e_0/B} - 1\right).$$

(2.32)

Solving equation (2.20), we get

$$e_1(t) = \alpha e^{-ct},$$

(2.33)

$$e_2(t) = -c\alpha e^{-ct}$$

(2.34)

where α is a constant. After determining α from relations (2.31) or (2.32), the above equations can be presented as

$$e_1(t) = \frac{B}{c^2}\left(-1 + e^{e_0 c^2/B}\right) e^{-ct},$$

(2.35)

$$e_2(t) = \frac{B}{c}\left(1 - e^{e_0 c^2/B}\right) e^{-ct}.$$

(2.36)

These two equations describe the tracking error for any time $t \geq t_f$. Notice that the error described by equations (2.24) and (2.35) does not exhibit any overshoots. This can be clearly seen from equations (2.25) and (2.36) which, taking into account inequality (2.13) and relation (2.28), demonstrate that the sign of the tracking error derivative $e_2(t)$ is always opposite to the sign of the initial error e_0. This implies that the error in the considered system dies out without oscillations or overshoots. Moreover, notice that neither external disturbance $d(t)$, nor the plant model uncertainty $\Delta f(x, t)$ appears in equations (2.24), (2.25), (2.35) and (2.36). This implies that the control law proposed in this chapter indeed ensures insensitivity of the system with respect to these signals from the very beginning of the control process. Now, in order to facilitate the procedure of determining the switching line parameters A, B and c, we introduce the following constant

$$m = \frac{e_0 c^2}{B} \tag{2.37}$$

which is always strictly positive because, as stated by relation (2.28), the signs of B and e_0 are always the same. With the above notation formulae (2.35) and (2.36) can be written more concisely as follows

$$e_1(t) = \frac{B}{c^2}\left(-1 + e^m\right)e^{-ct}, \tag{2.38}$$

$$e_2(t) = \frac{B}{c}\left(1 - e^m\right)e^{-ct}. \tag{2.39}$$

Our considerations presented up to now, as well as equations describing the system error evolution, clearly show that the switching line selection fully determines the system motion and its performance. Therefore, the switching line parameters A, B and c should be carefully chosen according to the user requirements and specifications, so as to obtain the best possible control quality without violating technical and environmental constraints. Therefore, in the next part of this chapter several methods of the switching line selection will be proposed. The methods ensure optimal, in the sense of either integral absolute error (IAE) or integral time multiplied absolute error (ITAE), performance of the system without exceeding the maximum admissible values of the input signal and/or the system error derivative. In any real system, where the demand output signal rate of change is limited, the latter constraint is directly related to the constraint imposed on the second state variable x_2.

In this book we will consider not only conventional constraints expressed by inequalities, but we will also take into account elastic constraints, which in many applications, may turn out to be more practical. The idea of these constraints is to allow for slightly higher (than the threshold) values of the considered signals if this brings significant improvement of the system performance. This type of constraints may be used in all non-critical applications, i.e. in those situations where using slightly bigger signals than the threshold value will not cause danger or put

2.2 Switching Line Design Minimising IAE

the safety of the operation at risk, but will merely result, in for example, a minor increase of fuel consumption.

2.2 Switching Line Design Minimising IAE

In this section the switching line parameters are selected. For that purpose the integral of the absolute error (IAE) is minimised subject to the input signal and velocity constraints. First we take into account each of the two constraints separately, and then we require both of them to be satisfied simultaneously. We begin with calculating the IAE

$$J_{IAE} = \int_{t_0}^{\infty} |e_1(t)| dt . \tag{2.40}$$

As we have demonstrated that the tracking error converges monotonically in the considered system, criterion (2.40) is equivalent to

$$J_{IAE} = \left| \int_0^{\infty} e_1(t) dt \right| . \tag{2.41}$$

Substituting equations (2.24) and (2.38) into (2.41), we obtain the following relation

$$J_{IAE} = \left| \int_0^{t_f} e_1(t) dt + \int_{t_f}^{\infty} e_1(t) dt \right| =$$

$$= \left| \int_0^{\frac{e_0 c}{B}} \left(-\frac{B}{c^2} e^{-ct} + e_0 + \frac{B}{c^2} - \frac{B}{c} t \right) dt + \int_{\frac{e_0 c}{B}}^{\infty} \left[\frac{B}{c^2} \left(-1 + e^m \right) e^{-ct} \right] dt \right| . \tag{2.42}$$

Then calculating appropriate integrals

$$J_{IAE} = \left| \left[\frac{B}{c^3} e^{-ct} + e_0 t + \frac{B}{c^2} t - \frac{B}{2c} t^2 \right]_0^{\frac{e_0 c}{B}} + \left[-\frac{B}{c^3} \left(-1 + e^m \right) e^{-ct} \right]_{\frac{e_0 c}{B}}^{\infty} \right| , \tag{2.43}$$

we get

$$J_{IAE} = \frac{e_0^2 c}{2|B|} + \frac{|e_0|}{c} . \tag{2.44}$$

On the other hand, calculating c from equation (2.37), we obtain

$$c = \sqrt{\frac{mB}{e_0}} . \tag{2.45}$$

In order to facilitate the minimisation process, further in the chapter we will consider the quality criterion as a function of variables m and B, instead of c and B. Substituting (2.45) into (2.44), the control quality criterion can be presented as follows

$$J_{\text{IAE}}\left(m,B\right)=\frac{\left|e_0\right|^{3/2}}{\sqrt{\left|B\right|}}\left(\frac{\sqrt{m}}{2}+\frac{1}{\sqrt{m}}\right).\tag{2.46}$$

Next, this criterion will be minimised, in order to determine the switching line parameters.

2.2.1 Switching Line Design Subject to Input Signal Constraint

Now we consider the situation when $u_{\max}$ is the maximum admissible value of the input signal. It means that we require that the following inequality holds

$$\left|u\left(t\right)\right|\leq u_{\max}\tag{2.47}$$

where $u_{\max}$ is a constant, which satisfies the following condition

$$u_{\max}>\frac{\left|\dot{x}_{2d}-f\left(x,t\right)\right|+\gamma}{\left|b\left(x,t\right)\right|}.\tag{2.48}$$

Conditions (2.47) and (2.48) imply that there exists such a strictly positive constant

$$U=u_{\max}-\max\left[\frac{\left|\dot{x}_{2d}-f\left(x,t\right)\right|+\gamma}{\left|b\left(x,t\right)\right|}\right]\tag{2.49}$$

that condition (2.47) is satisfied if the following relation holds

$$\left|\dot{e}_2\left(t\right)\right|\leq\left|b\left(x,t\right)\right|U.\tag{2.50}$$

Notice that from relations (2.11), (2.15), (2.49) and (2.50) we have

$$\left|u\left(t\right)\right|\leq\frac{1}{\left|b\left(x,t\right)\right|}\left[\left|\dot{e}_2\right|+\left|\dot{x}_{2d}-f\left(x,t\right)\right|+\gamma\right]\leq$$

$$\leq\frac{1}{\left|b\left(x,t\right)\right|}\left\{\left|b\left(x,t\right)\right|U+\left[\left|\dot{x}_{2d}-f\left(x,t\right)\right|+\gamma\right]\right\}=$$

$$=U+\frac{1}{\left|b\left(x,t\right)\right|}\left[\left|\dot{x}_{2d}-f\left(x,t\right)\right|+\gamma\right]\leq U+\max\left[\frac{\left|\dot{x}_{2d}-f\left(x,t\right)\right|+\gamma}{\left|b\left(x,t\right)\right|}\right]=$$

$$=u_{\max}-\max\left[\frac{\left|\dot{x}_{2d}-f\left(x,t\right)\right|+\gamma}{\left|b\left(x,t\right)\right|}\right]+\max\left[\frac{\left|\dot{x}_{2d}-f\left(x,t\right)\right|+\gamma}{\left|b\left(x,t\right)\right|}\right]=u_{\max}.$$

$$\tag{2.51}$$

2.2 Switching Line Design Minimising IAE

This inequality proves that the constant U specified above actually exists. Furthermore, clearly inequality (2.50) always holds if

$$\left|\dot{e}_2\left(t\right)\right|\leq\delta U .\tag{2.52}$$

Let us calculate the greatest value of $\left|\dot{e}_2(t)\right|$. For $t\leq t_f$ the tracking error is given by equation (2.24). Differentiating this equation two times – or alternatively differentiating equation (2.25) once – we get

$$\dot{e}_2\left(t\right)=-Be^{-ct} .\tag{2.53}$$

The maximum absolute value of this signal, achieved at the initial time $t_0=0$ is equal to $\left|\dot{e}_2(0)\right|=\left|B\right|$. On the other hand, for $t\geq t_f$ the tracking error is described by equation (2.38) whose second derivative right hand side absolute value decreases with time. Consequently, for $t\geq t_f$ the maximum value of $\left|\dot{e}_2(t)\right|$, reached at the time t_f, equals

$$\begin{aligned}\left|\dot{e}_2\left(t_f\right)\right|&=\left|-B\left(1-e^m\right)e^{-ct_f}\right|=\left|-B\left(1-e^m\right)e^{-m}\right|=\\&=\left|-B\left(e^{-m}-1\right)\right|=\left|B\left(1-e^{-m}\right)\right|.\end{aligned}\tag{2.54}$$

As m is greater than zero, the right hand side of the above equation is always smaller than $|B|$. Hence, we conclude that $|B|$ is the extreme value of $\left|\dot{e}_2(t)\right|$, and from relation (2.52) we obtain the following constraint

$$\left|B\right|\leq\delta U .\tag{2.55}$$

Now we will precisely analyse the criterion J_{IAE} minimisation task. Notice that for any given value of m, the minimum of criterion (2.46) is obtained for the greatest value of $|B|$ satisfying constraint (2.55). Therefore, the solution of the considered minimisation task can be found as a minimum of a single variable function $J_{IAE}(m)$. Taking into account constraint (2.55) we get

$$J_{IAE}^u\left(m\right)=\frac{\left|e_0\right|^{3/2}}{\sqrt{\delta U}}\left(\frac{\sqrt{m}}{2}+\frac{1}{\sqrt{m}}\right).\tag{2.56}$$

In order to analyse the minimisation task we calculate the derivative of expression (2.56) with respect to m

$$\frac{dJ_{IAE}^u\left(m\right)}{dm}=\frac{\left|e_0\right|^{3/2}}{\sqrt{\delta U}}\left(\frac{1}{4\sqrt{m}}-\frac{1}{2m\sqrt{m}}\right).\tag{2.57}$$

Equating this derivative to zero we obtain the following optimal solution $m_{u\,opt}=2$, which minimises criterion (2.56). Then for that value of m, from relations (2.28) and (2.55), we get the optimal value of parameter B

$$B_{u\,opt}=\delta U\,\mathrm{sgn}\left(e_0\right).\tag{2.58}$$

This concludes the minimisation of criterion (2.46) subject to constraint (2.47). The other optimal switching line parameters, calculated from relations (2.45) and (2.23), are expressed by the following formulae

$$c_{u\ opt} = \sqrt{\frac{2\ \delta U\ \mathrm{sgn}(e_0)}{e_0}} = \sqrt{\frac{2\ \delta U}{|e_0|}}, \qquad (2.59)$$

$$A_{u\ opt} = -\sqrt{\frac{2\ \delta U}{|e_0|}}e_0 = -\mathrm{sgn}(e_0)\sqrt{2\ \delta U |e_0|}. \qquad (2.60)$$

The switching line stops moving at the time instant

$$t_{f\ u\ opt} = \sqrt{\frac{2|e_0|}{\delta U}}. \qquad (2.61)$$

The minimum value of criterion (2.56) is equal to

$$J^u_{\mathrm{IAE}}(2) = \frac{\sqrt{2}|e_0|^{3/2}}{\sqrt{\delta U}}. \qquad (2.62)$$

In the circumstances considered in this section the switching line is shifted along e_1 axis at the rate $-\sqrt{\delta U |e_0|/2}\,\mathrm{sgn}(e_0)$. This motion of the line can also be interpreted as a vertical translation along e_2 axis with velocity equal to $-B_{u\ opt} = -\delta U\mathrm{sgn}(e_0)$.

In the next part of this section, another method of the switching line design will be presented. The method will ensure good dynamic performance of the considered system subject to elastic input constraint. In fact, in this method we minimise a modified criterion which is a sum of the IAE and a penalty function for using the control signal of excessive magnitude. The criterion can be expressed as follows

$$Q^u_{\mathrm{IAE}} = \int_{t_0}^{\infty}|e_1(t)|dt + q\left\{\frac{\max\left[|u(t)|\right]}{M}\right\}^n \qquad (2.63)$$

where $M > 0$ is the threshold value of the input signal, $q > 0$ is a weighting factor and $n \geq 1$ is a constant determining how elastic (or stretchable) the constraint is. This criterion determines the regulation quality in the system and at the same time penalises excessive values of $|u(t)|$. Therefore, it can be used to design efficient control systems with soft input constraints. On the other hand, putting $n \to \infty$, choosing M equal to the maximum admissible value of $|u(t)|$ and minimising criterion (2.63) we will design a control law which guarantees the minimum integral of the absolute error (IAE) strictly without exceeding the maximum admissible value of input signal, i.e. we will obtain the result presented in the previous part of this section as a limit case of the situation considered here. Taking into account the fact that the tracking error converges to zero monotonically and considering relations (2.40), (2.44) and (2.46), criterion (2.63) can be presented as

2.2 Switching Line Design Minimising IAE

$$Q_{IAE}^{u}\left(m,|B|\right)=\frac{|e_{0}|^{3/2}}{\sqrt{|B|}}\left(\frac{\sqrt{m}}{2}+\frac{1}{\sqrt{m}}\right)+q\left\{\frac{\max\left[|u(t)|\right]}{M}\right\}^{n}. \tag{2.64}$$

Since u_{max} is the threshold value of the input signal we get

$$Q_{IAE}^{u}\left(m,|B|\right)=\frac{|e_{0}|^{3/2}}{\sqrt{|B|}}\left(\frac{\sqrt{m}}{2}+\frac{1}{\sqrt{m}}\right)+q\left\{\frac{\max\left[|u(t)|\right]}{u_{max}}\right\}^{n}. \tag{2.65}$$

Our previous considerations show that for the purpose of the sliding line design subject to input signal constraint, criterion (2.65) can be replaced with

$$Q_{IAE}^{u}\left(m,|B|\right)=\frac{|e_{0}|^{3/2}}{\sqrt{|B|}}\left(\frac{\sqrt{m}}{2}+\frac{1}{\sqrt{m}}\right)+q\left\{\frac{\max\left[|\dot{e}_{2}(t)|\right]}{\delta U}\right\}^{n}. \tag{2.66}$$

Now we will analyse the minimisation task of criterion (2.66). Since we have already demonstrated that $\max\left[|\dot{e}_{2}(t)|\right]$ is expressed by $|B|$, criterion (2.66) can be reformulated as

$$Q_{IAE}^{u}\left(m,|B|\right)=\frac{|e_{0}|^{3/2}}{\sqrt{|B|}}\left(\frac{\sqrt{m}}{2}+\frac{1}{\sqrt{m}}\right)+q\left(\frac{|B|}{\delta U}\right)^{n}. \tag{2.67}$$

Now this criterion will be minimised as a function of two variables: m and $|B|$. The criterion can be minimised by finding its partial derivatives with respect to both variables and equating these derivatives to zero. The partial derivatives have the following form

$$\frac{\partial Q_{IAE}^{u}\left(m,|B|\right)}{\partial m}=\frac{|e_{0}|^{3/2}}{\sqrt{|B|}}\left(\frac{1}{4\sqrt{m}}-\frac{1}{2m\sqrt{m}}\right), \tag{2.68}$$

$$\frac{\partial Q_{IAE}^{u}\left(m,|B|\right)}{\partial|B|}=\frac{-|e_{0}|^{3/2}}{2|B|\sqrt{|B|}}\left(\frac{\sqrt{m}}{2}+\frac{1}{\sqrt{m}}\right)+nq\frac{|B|^{n-1}}{\left(\delta U\right)^{n}}. \tag{2.69}$$

Notice that solving equation

$$\frac{\partial Q_{IAE}^{u}\left(m,|B|\right)}{\partial m}=0, \tag{2.70}$$

we get $m=2$. On the other hand, from

$$\frac{\partial Q_{IAE}^{u}\left(m,|B|\right)}{\partial|B|}=0 \tag{2.71}$$

we obtain the following relation

$$|B|^{n+\frac{1}{2}} = \frac{|e_0|^{3/2}(\delta U)^n}{2qn}\left(\frac{\sqrt{m}}{2}+\frac{1}{\sqrt{m}}\right). \tag{2.72}$$

Therefore, we expect that criterion (2.67) achieves its minimum value for $m = 2$ and the following value of parameter $|B|$

$$|B_0| = \left[\frac{\sqrt{2}|e_0|^{3/2}(\delta U)^n}{2qn}\right]^{\frac{2}{2n+1}}. \tag{2.73}$$

In order to verify this property, we calculate the second order derivatives of criterion Q_{IAE}^u

$$\frac{\partial^2 Q_{\text{IAE}}^u(m,|B|)}{\partial m \partial m} = \frac{|e_0|^{3/2}}{\sqrt{|B|}}\left(-\frac{1}{8m\sqrt{m}}+\frac{3}{4m^2\sqrt{m}}\right), \tag{2.74}$$

$$\frac{\partial^2 Q_{\text{IAE}}^u(m,|B|)}{\partial m \partial|B|} = \frac{\partial^2 Q_{\text{IAE}}^u(m,|B|)}{\partial|B|\partial m} = -\frac{|e_0|^{3/2}}{2|B|\sqrt{|B|}}\left(\frac{1}{4\sqrt{m}}-\frac{1}{2m\sqrt{m}}\right), \tag{2.75}$$

$$\frac{\partial^2 Q_{\text{IAE}}^u(m,|B|)}{\partial|B|\partial|B|} = \frac{3|e_0|^{3/2}}{4|B|^2\sqrt{|B|}}\left(\frac{\sqrt{m}}{2}+\frac{1}{\sqrt{m}}\right)+qn(n-1)\frac{|B|^{n-2}}{(\delta U)^n}. \tag{2.76}$$

Then, for $m = 2$ and $|B| = |B_0|$ we build the matrix of the second order derivatives

$$H = \begin{bmatrix} \dfrac{\partial^2 Q_{\text{IAE}}^u(m,|B|)}{\partial m \partial m}\bigg|_{\substack{m=2\\|B|=|B_0|}} & \dfrac{\partial^2 Q_{\text{IAE}}^u(m,|B|)}{\partial m \partial|B|}\bigg|_{\substack{m=2\\|B|=|B_0|}} \\[4ex] \dfrac{\partial^2 Q_{\text{IAE}}^u(m,|B|)}{\partial|B|\partial m}\bigg|_{\substack{m=2\\|B|=|B_0|}} & \dfrac{\partial^2 Q_{\text{IAE}}^u(m,|B|)}{\partial|B|\partial|B|}\bigg|_{\substack{m=2\\|B|=|B_0|}} \end{bmatrix}. \tag{2.77}$$

Substituting relations (2.74), (2.75) and (2.76) into (2.77), for $m = 2$ and $|B| = |B_0|$ we obtain

$$H = \begin{bmatrix} \dfrac{1}{8}\left(\dfrac{|e_0|^{3/2}}{\sqrt{2}}\right)^{\frac{2n}{2n+1}}\left[\dfrac{qn}{(\delta U)^n}\right]^{\frac{1}{2n+1}} & 0 \\[4ex] 0 & \left(\dfrac{|e_0|^{3/2}}{\sqrt{2}}\right)^{\frac{2n-4}{2n+1}}\left[\dfrac{qn}{(\delta U)^n}\right]^{\frac{5}{2n+1}}\left(n+\dfrac{1}{2}\right) \end{bmatrix}. \tag{2.78}$$

Notice that the following two inequalities are always satisfied

2.2 Switching Line Design Minimising IAE

$$\frac{1}{8}\left(\frac{|e_0|^{3/2}}{\sqrt{2}}\right)^{\frac{2n}{2n+1}}\left[\frac{qn}{(\delta U)^n}\right]^{\frac{1}{2n+1}} > 0, \tag{2.79}$$

$$\left(\frac{|e_0|^{3/2}}{\sqrt{2}}\right)^{\frac{2n-4}{2n+1}}\left[\frac{qn}{(\delta U)^n}\right]^{\frac{5}{2n+1}}\left(n+\frac{1}{2}\right) > 0. \tag{2.80}$$

This leads us to the conclusion that $\det(H) > 0$ and matrix H is positive definite. Hence, for $m = 2$ and

$$B_{u\,opt} = B_0 = \left[\frac{\sqrt{2}|e_0|^{3/2}(\delta U)^n}{2qn}\right]^{\frac{2}{2n+1}}\mathrm{sgn}(e_0) \tag{2.81}$$

the considered criterion reaches its minimum. Thus we conclude that $m = 2$ and $B_{u\,opt}$ described by (2.81) are the optimal (in the sense of criterion (2.66)) parameters of the considered switching line. The other two parameters of the line are given as follows

$$c_{u\,opt} = \left[\frac{|e_0|^{3/2}(\delta U)^n}{\sqrt{2}qn}\right]^{\frac{1}{2n+1}}\sqrt{\frac{2}{|e_0|}}, \tag{2.82}$$

$$A_{u\,opt} = -e_0\left[\frac{|e_0|^{3/2}(\delta U)^n}{\sqrt{2}qn}\right]^{\frac{1}{2n+1}}\sqrt{\frac{2}{|e_0|}} = -\mathrm{sgn}(e_0)\left[\frac{|e_0|^{3/2}(\delta U)^n}{\sqrt{2}qn}\right]^{\frac{1}{2n+1}}\sqrt{2|e_0|}. \tag{2.83}$$

The line stops moving at the time

$$t_{f\,u\,opt} = \sqrt{2|e_0|}\left[\frac{\sqrt{2}qn}{|e_0|^{3/2}(\delta U)^n}\right]^{\frac{1}{2n+1}}. \tag{2.84}$$

Notice that putting $n \to \infty$ into above equations and calculating appropriate limits of (2.81), (2.82), (2.83) and (2.84) we get the switching line parameters expressed by (2.58), (2.59), (2.60) and (2.61), respectively. These parameters ensure the optimal, in the sense of the minimum integral absolute error, performance of system (2.1) subject to the conventional (i.e. non-elastic) input constraint $|u(t)| \le u_{max}$. This shows that the conventional constraint considered at the beginning of this section is a special case of the elastic constraint analysed here. Criterion Q_{IAE}^u as a function of two variables m and $|B|$ is illustrated in figure 2.2.

Fig. 2.2 Criterion $Q_{IAE}^u(m,|B|)$

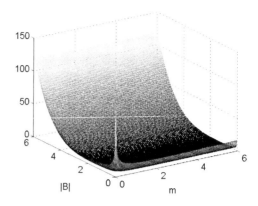

Simulation example

In order to verify the proposed method of the switching line design we consider the following second order system

$$\dot{x}_1 = x_2$$
$$\dot{x}_2 = (1+\varepsilon)\sin x_1 + \frac{1}{2}-\exp(-x_2^2)+u+d(t) \qquad (2.85)$$

where

$$d(t) = 0.6\sin(10t) \qquad (2.86)$$

is the external disturbance and $\varepsilon = 0.39$ represents the model uncertainty. Consequently we select $\gamma = 1$. The initial conditions

$$x_{10} = -\pi/2, \; x_{20} = 0 \qquad (2.87)$$

and the demand trajectory is given as

$$x_{1d}(t) = -\cos t. \qquad (2.88)$$

The input signal is required not to exceed its admissible value $u_{max} = 5$. From this condition, taking into account relations (2.49), (2.86) and (2.88), we get $U = 1.5$. Calculating the parameters of the switching line according to the proposed method, i.e. using formulae (2.58), (2.59) and (2.60) we obtain the following optimal coefficients $B_{u\,opt} = -1.5$, $c_{u\,opt} \approx 2.29$, $A_{u\,opt} \approx 1.31$. The considered line stops moving at the time instant $t_{fu\,opt}$ equal to 0.87. Figure 2.3 shows evolution of the tracking error and its derivative in the considered system. It can be seen from this figure that the tracking error converges to zero monotonically and furthermore, that the system is insensitive with respect to the external disturbance and the model uncertainty from the very beginning of the control action. Figure 2.4 shows the input signal and demonstrates that the constraint of this signal $|u(t)| \leq u_{max} = 5$ is always

2.2 Switching Line Design Minimising IAE

satisfied. Furthermore, figures 2.5 – 2.8 illustrate the performance of the same system in the situation when, in the sliding line design process, elastic input signal constraint is taken into account. The plots obtained for $n = 5$ and $q = 5$ are presented in figures 2.5 and 2.6. Then for the purpose of comparison, in figures 2.7 and 2.8 we show the same plots but obtained for $n = 5$ and $q = 0.02$. It can be seen from the figures how different selection of weighting factor q affects the system performance and the magnitude of its input signal. Decreasing q results in faster error convergence, however this desirable effect is obtained at the expense of rising magnitude of the input signal. Finally, figure 2.9 shows phase trajectories of system (2.85) with the same three controllers. Curve 1 in this figure represents the phase trajectory obtained when conventional input constraint is considered, and curves 2 and 3 show the same trajectory but generated when elastic constraint is taken into account. Curve 2 demonstrates the system performance when $n = 5$ and $q = 5$ and plot 3 presents the system behaviour when $n = 5$ and $q = 0.02$.

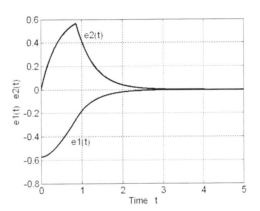

Fig. 2.3 Tracking error and its derivative (conventional input constraint)

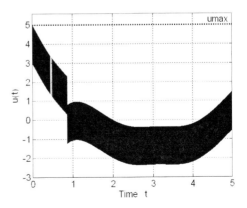

Fig. 2.4 Control signal (conventional input constraint)

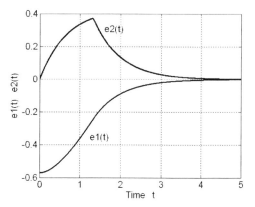

Fig. 2.5 Tracking error and its derivative ($n = 5$, $q = 5$)

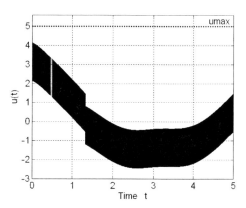

Fig. 2.6 Control signal ($n = 5$, $q = 5$)

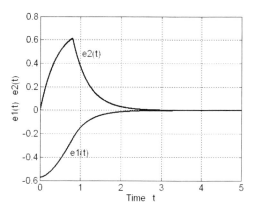

Fig. 2.7 Tracking error and its derivative ($n = 5$, $q = 0.02$)

Fig. 2.8 Control signal ($n = 5$, $q = 0.02$)

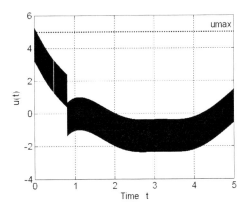

Fig. 2.9 Phase trajectories

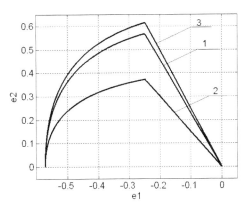

2.2.2 Switching Line Design Subject to Velocity Constraint

In this section we will consider system (2.1) subject to the velocity constraint

$$|e_2(t)| \leq v_{max} \tag{2.89}$$

where v_{max} is the maximum admissible velocity of the considered system. For any time $t \leq t_f$ the system velocity is described by equation (2.25) and for the time $t \geq t_f$ by equation (2.36). Since the absolute value of the right hand side of equation (2.25) represents an increasing function of time and the absolute value of the right hand side of equation (2.36) is a decreasing function of the same argument, the extreme value of the velocity is achieved at the time instant t_f. This extreme value

$$e_2(t_f) = \frac{B}{c}(e^{-m} - 1). \tag{2.90}$$

Thus, in order to satisfy constraint (2.89), we require that

$$\left|\frac{B}{c}(e^{-m}-1)\right| \leq v_{max}. \tag{2.91}$$

From this inequality, using relation (2.45), we find the maximum admissible value of $|B|$

$$|B| \leq \left[\frac{v_{max}\sqrt{m}}{\sqrt{|e_0|}(1-e^{-m})}\right]^2 \tag{2.92}$$

and substituting this value into expression (2.46), we get

$$J_{IAE}^v(m) = \frac{e_0^2}{v_{max}}(1-e^{-m})\left(\frac{1}{2}+\frac{1}{m}\right). \tag{2.93}$$

This function, for any fixed m expresses the minimum value of criterion $J_{IAE}(m, B)$ which can be achieved when velocity constraint (2.89) is satisfied. As shown in figure 2.10, $J_{IAE}^v(m)$ is a decreasing function of its argument m, and therefore in this case, i.e. when the velocity constraint is taken into account, we get $m_{v\,opt} \to \infty$. Since parameter B can be calculated substituting $m_{v\,opt}$ into

Fig. 2.10 Criterion $J_{IAE}^v(m)$

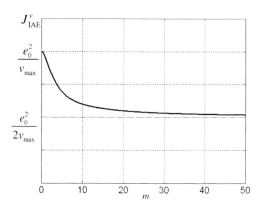

$$B = \left[\frac{v_{max}\sqrt{m}}{\sqrt{|e_0|}(1-e^{-m})}\right]^2 \operatorname{sgn}(e_0), \tag{2.94}$$

then

$$B_{v\,opt} \to \operatorname{sgn}(e_0)\cdot\infty. \tag{2.95}$$

Consequently, from relations (2.45), (2.23) and (2.27) we obtain

$$c_{v\,opt} \to \infty, \tag{2.96}$$

$$A_{v\,opt} \to -\operatorname{sgn}(e_0)\cdot\infty \tag{2.97}$$

2.2 Switching Line Design Minimising IAE

and

$$t_{f\,v\,opt} = \frac{|e_0|}{v_{max}}.$$ (2.98)

These formulae imply that in the situation considered in this section, the switching line is always vertical and it moves along the horizontal axis of the (e_1, e_2) coordinate frame at the rate $- v_{max} \, \text{sgn}(e_0)$.

Now, as we did in the previous subsection, we modify the considered criterion so that we can introduce an elastic velocity constraint which allows the threshold value of the system velocity to be slightly exceeded. This modified criterion has the following form

$$Q_{IAE}^v = \int_{t_0}^{\infty} |e_1(t)| dt + q \left\{ \frac{\max\left[|e_2(t)|\right]}{M} \right\}^n$$ (2.99)

where now $M > 0$ is the threshold value of the system velocity, $q > 0$ is a weighting factor and $n \geq 1$ is a constant determining how elastic the constraint is. Taking into account our previous considerations given by (2.40) – (2.46) and the assumption that v_{max} is the threshold value of the system velocity we get

$$Q_{IAE}^v = \frac{|e_0|^{3/2}}{\sqrt{|B|}} \left(\frac{\sqrt{m}}{2} + \frac{1}{\sqrt{m}} \right) + q \left\{ \frac{\max\left[|e_2(t)|\right]}{v_{max}} \right\}^n.$$ (2.100)

Now we minimise criterion (2.100). Because $\max[|e_2(t)|]$ is described by the absolute value of (2.90), taking into account (2.45) we can express the considered criterion as

$$Q_{IAE}^v = \frac{|e_0|^{3/2}}{\sqrt{|B|}} \left(\frac{\sqrt{m}}{2} + \frac{1}{\sqrt{m}} \right) + q \left[\frac{\sqrt{|B|}|e_0|\left(1-e^{-m}\right)}{v_{max}\,\sqrt{m}} \right]^n.$$ (2.101)

Notice that for any m, derivative

$$\frac{\partial Q_{IAE}^v}{\partial |B|} = -\frac{|e_0|^{3/2}}{2|B|\sqrt{|B|}} \left(\frac{\sqrt{m}}{2} + \frac{1}{\sqrt{m}} \right) + \frac{1}{2} qn|B|^{(n-2)/2} \left[\frac{\sqrt{|e_0|}\left(1-e^{-m}\right)}{v_{max}\,\sqrt{m}} \right]^n$$ (2.102)

is negative for small values of $|B|$, and for greater values of this parameter it becomes positive. Solving equation

$$\frac{\partial Q_{IAE}^v}{\partial |B|} = 0,$$ (2.103)

we obtain

$$|B| = \left[\frac{|e_0|^{3/2} \left(\dfrac{\sqrt{m}}{2} + \dfrac{1}{\sqrt{m}} \right) \left(v_{\max} \sqrt{m} \right)^n}{qn \left(\sqrt{|e_0|} \right)^n \left(1 - e^{-m} \right)^n} \right]^{\frac{2}{n+1}}. \tag{2.104}$$

Thus for any fixed m, criterion Q_{IAE}^v reaches its minimum for $|B|$ given by expression (2.104). Then substituting expression (2.104) into criterion (2.101) we obtain

$$Q_{\mathrm{IAE}}^v = \frac{|e_0|^{3/2} \left[qn \left(\sqrt{|e_0|} \right)^n \left(1 - e^{-m} \right)^n \right]^{\frac{1}{n+1}}}{\left[|e_0|^{3/2} \left(\dfrac{\sqrt{m}}{2} + \dfrac{1}{\sqrt{m}} \right) \left(v_{\max} \sqrt{m} \right)^n \right]^{\frac{1}{n+1}}} \left(\frac{\sqrt{m}}{2} + \frac{1}{\sqrt{m}} \right) +$$

$$+ q \left\{ \frac{\left[\left[|e_0|^{3/2} \left(\dfrac{\sqrt{m}}{2} + \dfrac{1}{\sqrt{m}} \right) \left(v_{\max} \sqrt{m} \right)^n \right]^{\frac{1}{n+1}} \sqrt{|e_0|} \left(1 - e^{-m} \right) \right]^n}{\left[qn \left(\sqrt{|e_0|} \right)^n \left(1 - e^{-m} \right)^n \right]^{\frac{1}{n+1}} v_{\max} \sqrt{m}} \right\}. \tag{2.105}$$

From this formula after some calculations we get

$$Q_{\mathrm{IAE}}^v = \frac{\left(|e_0|^{3/2} \right)^{\frac{n}{n+1}} \left[qn \left(\sqrt{|e_0|} \right)^n \left(1 - e^{-m} \right)^n \right]^{\frac{1}{n+1}}}{\left(v_{\max} \sqrt{m} \right)^{\frac{n}{n+1}}} \left(\frac{\sqrt{m}}{2} + \frac{1}{\sqrt{m}} \right)^{\frac{n}{n+1}} +$$

$$+ q \frac{\left[|e_0|^{3/2} \left(\dfrac{\sqrt{m}}{2} + \dfrac{1}{\sqrt{m}} \right) \left(v_{\max} \sqrt{m} \right)^n \right]^{\frac{n}{n+1}} \left[\sqrt{|e_0|} \left(1 - e^{-m} \right) \right]^{\frac{n}{n+1}}}{\left(qn \right)^{\frac{n}{n+1}} \left(v_{\max} \sqrt{m} \right)^n} = \tag{2.106}$$

$$= \frac{\left(e_0^2 \right)^{\frac{n}{n+1}} \left(qn \right)^{\frac{1}{n+1}} \left(1 - e^{-m} \right)^{\frac{n}{n+1}}}{\left(v_{\max} \right)^{\frac{n}{n+1}}} \left(\frac{1}{2} + \frac{1}{m} \right)^{\frac{n}{n+1}} +$$

$$+ q \frac{\left(e_0^2 \right)^{\frac{n}{n+1}} \left(1 - e^{-m} \right)^{\frac{n}{n+1}}}{\left(qn \right)^{\frac{n}{n+1}} \left(v_{\max} \right)^{\frac{n}{n+1}}} \left(\frac{1}{2} + \frac{1}{m} \right)^{\frac{n}{n+1}},$$

2.2 Switching Line Design Minimising IAE

and finally

$$Q_{IAE}^v = (qn)^{\frac{1}{n+1}} \left[e_0^2 \left(\frac{1}{2} + \frac{1}{m} \right) \frac{(1-e^{-m})}{v_{max}} \right]^{\frac{n}{n+1}} \left(1 + \frac{1}{n} \right). \qquad (2.107)$$

Taking into account evolution of function (2.93) shown in figure 2.10, we conclude that criterion (2.107) achieves its minimum, when m tends to infinity. Then substituting $m_{v\,opt} \to \infty$ into relation (2.104) we find that the optimal value of parameter $|B|$ also tends to infinity. Then, the set of the optimal parameters is given by formulae (2.95) – (2.97) and

$$t_{f\,v\,opt} = \left(\frac{2qn|e_0|^{n-1}}{v_{max}^n} \right)^{\frac{1}{n+1}}. \qquad (2.108)$$

In this case the vertical switching line moves horizontally along e_1 axis and approaches the origin of the (e_1, e_2) coordinate frame, with the constant velocity $-|e_0|^{2/(n+1)} v_{max}^{n/(n+1)} (2qn)^{-1/(n+1)} \mathrm{sgn}(e_0)$. Notice that when n tends to infinity, this velocity approaches $-v_{max}\mathrm{sgn}(e_0)$ and the time when the line stops moving is the same as the time obtained in the case of conventional constraint. Criterion Q_{IAE}^v as a function of two variables m and $|B|$ is presented in figure 2.11.

Fig. 2.11 Criterion $Q_{IAE}^v(m,|B|)$

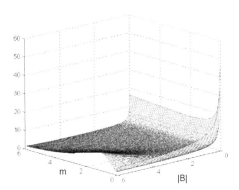

That concludes the analysis of the elastic velocity constraint.

Simulation example

In order to verify the proposed control method we again consider system (2.85) with the external disturbance given by (2.86) and γ equal to 1. The initial conditions of the system are given by (2.87) and the demand trajectory is defined by (2.88). The maximum admissible value of the system velocity v_{max} is equal to 0.5.

Taking into account the conventional velocity constraint, according to the above method, we obtain that $B_{v\,opt} \to -\infty$, $c_{v\,opt} \to \infty$, $A_{v\,opt} \to \infty$ and $t_{fv\,opt} \approx 1.142$. Figures 2.12 and 2.13 show the tracking error evolution and its derivative in this case. Then figure 2.14 demonstrates the system phase trajectory in the error state space. The input signal is presented in the figure 2.15. It can be seen from the figures that this control strategy requires infinite control signal (unit impulses at $t = 0$ and $t \approx 1.142$) and therefore it is not a feasible option for the considered system. Thus, in the next section we will take into account both velocity and input signal constraints simultaneously. The next four figures 2.16 – 2.19 show the performance of the same system with elastic velocity constraint, when $n = 5$ and $q = 5$. Furthermore, figures 2.20 – 2.23 again present the behaviour of the system with elastic velocity constraint but this time with $n = 5$ and $q = 0.02$. It can be clearly seen from the figures how weights selection affects the system dynamics. Increasing parameter q as expected results in smaller velocity and the sluggish system response.

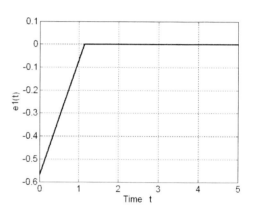

Fig. 2.12 Tracking error (conventional velocity constraint)

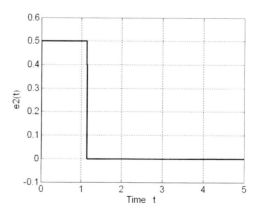

Fig. 2.13 Tracking error derivative (conventional velocity constraint)

2.2 Switching Line Design Minimising IAE

Fig. 2.14 Phase trajectory (conventional velocity constraint)

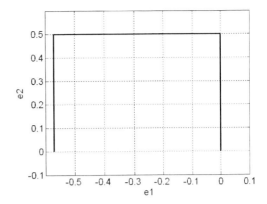

Fig. 2.15 Control signal (conventional velocity constraint)

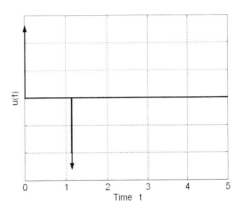

Fig. 2.16 Tracking error ($n = 5, q = 5$)

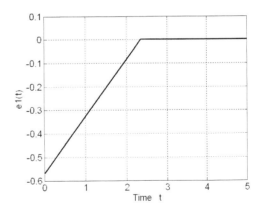

Fig. 2.17 Tracking error derivative ($n = 5$, $q = 5$)

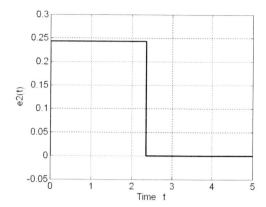

Fig. 2.18 Phase trajectory ($n = 5$, $q = 5$)

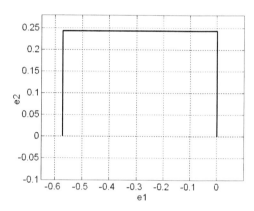

Fig. 2.19 Control signal ($n = 5$, $q = 5$)

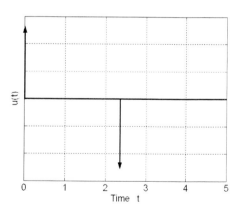

2.2 Switching Line Design Minimising IAE

Fig. 2.20 Tracking error ($n = 5$, $q = 0.02$)

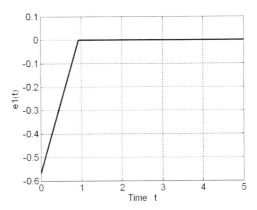

Fig. 2.21 Tracking error derivative ($n = 5$, $q = 0.02$)

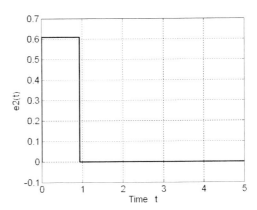

Fig. 2.22 Phase trajectory ($n = 5$, $q = 0.02$)

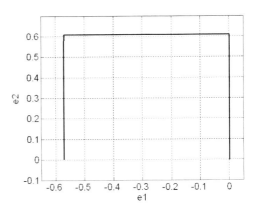

Fig. 2.23 Control signal ($n = 5$, $q = 0.02$)

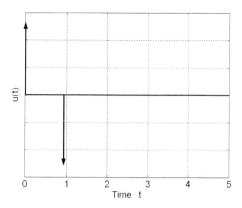

Up to now we determined the optimal switching line parameters when each of the constraints is taken into account separately. Further in this chapter, we move on to the case when more than one constraint has to be satisfied at the same time. Even though, in that situation again both conventional and elastic constraints could be considered, for sake of clarity, we will confine the following presentation to system (2.1) subject to conventional (non-stretchable) constraints only.

2.2.3 Switching Line Design Subject to Input Signal and Velocity Constraints

In this section we present the switching line design when u_{max} is the maximum admissible value of the input signal, and the velocity cannot exceed v_{max}, i.e. we take into account constraints (2.47) and (2.89), simultaneously. Therefore, in order to find the minimum value of criterion $J_{IAE}(m, B)$ expressed by (2.46) we will now minimise the following function

$$J_{IAE}^{uv}(m) = \max\left[J_{IAE}^{u}(m), J_{IAE}^{v}(m)\right]. \quad (2.109)$$

The optimal solution of the minimisation task will be such a value of m which satisfies the following condition

$$J_{IAE}^{uv}(m_{uv\,opt}) = \min_{m>0}\left\{\max\left[J_{IAE}^{u}(m), J_{IAE}^{v}(m)\right]\right\} \quad (2.110)$$

and the respective value of parameter B. For sake of clarity, we recall that functions $J_{IAE}^{u}(m)$ and $J_{IAE}^{v}(m)$ in equation (2.110) are determined by (2.56) and (2.93), respectively. First, we consider the following two cases: $J_{IAE}^{v}(2) \leq J_{IAE}^{u}(2)$ and $J_{IAE}^{v}(2) > J_{IAE}^{u}(2)$. Notice that in the first case, i.e. $J_{IAE}^{v}(2) \leq J_{IAE}^{u}(2)$, the input signal constraint actually limits the system performance and the velocity bound does not play a crucial role in the considered optimisation problem. Therefore, we get

2.2 Switching Line Design Minimising IAE

$$\min_{m>0}\left\{\max\left[J_{\mathrm{IAE}}^{u}\left(m\right), J_{\mathrm{IAE}}^{v}\left(m\right)\right]\right\} = J_{\mathrm{IAE}}^{u}\left(2\right). \tag{2.111}$$

Hence the optimal value of parameter m is $m_{uv\,opt} = m_{u\,opt} = 2$ and the optimal value $B_{uv\,opt} = B_{u\,opt}$ can be found from (2.58). Furthermore, the optimal parameters $c_{uv\,opt} = c_{u\,opt}$, $A_{uv\,opt} = A_{u\,opt}$, $t_{f\,uv\,opt} = t_{f\,u\,opt}$ can be determined from equations (2.59), (2.60) and (2.61), respectively.

Analysing the latter case $J_{\mathrm{IAE}}^{v}\left(2\right) > J_{\mathrm{IAE}}^{u}\left(2\right)$, one may notice that function $J_{\mathrm{IAE}}^{u}\left(m\right)$ increases for any m greater than 2 and its limit

$$\lim_{m\to\infty} J_{\mathrm{IAE}}^{u}\left(m\right) = \infty. \tag{2.112}$$

Furthermore, $J_{\mathrm{IAE}}^{v}\left(m\right)$ is a decreasing function of its argument in its whole domain. Moreover,

$$\lim_{m\to\infty} J_{\mathrm{IAE}}^{v}\left(m\right) = \frac{e_0^2}{2v_{max}} < \infty. \tag{2.113}$$

This implies that in this case the optimal value $m_{uv\,opt}$ is determined by this root of the following equation

$$J_{\mathrm{IAE}}^{u}\left(m_{uv\,opt}\right) = J_{\mathrm{IAE}}^{v}\left(m_{uv\,opt}\right) \tag{2.114}$$

which is greater than 2. Functions $J_{\mathrm{IAE}}^{u}\left(m\right)$ and $J_{\mathrm{IAE}}^{v}\left(m\right)$ are illustrated in figure 2.24.

We will show how to find the greater root of equation (2.114). Let us first notice that the following inequality holds for any positive m

$$J_{\mathrm{IAE}}^{v}\left(m\right) = \frac{e_0^2}{v_{max}}\left(\frac{1}{2} + \frac{1}{m}\right)\left(1 - e^{-m}\right) < \frac{e_0^2}{v_{max}}\left(\frac{1}{2} + \frac{1}{m}\right). \tag{2.115}$$

Therefore, the right hand side of (2.115) dominates criterion $J_{\mathrm{IAE}}^{v}\left(m\right)$. Thus solving equation

$$\frac{e_0^2}{v_{max}}\left(\frac{1}{2} + \frac{1}{m}\right) = J_{\mathrm{IAE}}^{u}\left(m\right) \tag{2.116}$$

which is equivalent to

$$\frac{e_0^2}{v_{max}}\left(\frac{1}{2} + \frac{1}{m}\right) = \frac{|e_0|^{3/2}\sqrt{m}}{\sqrt{\delta U}}\left(\frac{1}{2} + \frac{1}{m}\right), \tag{2.117}$$

we obtain such a value m of argument m that $m_{uv\,opt}$ is smaller than m. Actually

$$m_\gamma = \frac{|e_0|\delta U}{v_{max}^2}. \tag{2.118}$$

Fig. 2.24 Criteria $J^v_{IAE}(m)$ and $J^u_{IAE}(m)$

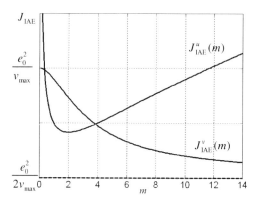

Since $m_{uv\,opt}$ is greater than 2 this implies that $m_{uv\,opt} \in (2, m_\gamma)$. In order to find $m_{uv\,opt}$ the root of the following function

$$f_1(m) = J^v_{IAE}(m) - J^u_{IAE}(m) = |e_0|^{3/2}\left(\frac{1}{2}+\frac{1}{m}\right)\left[\frac{\sqrt{|e_0|}(1-e^{-m})}{v_{max}} - \sqrt{\frac{m}{\delta U}}\right] \quad (2.119)$$

may be calculated numerically in the interval $(2, m_\gamma)$. Indeed in this interval $f_1(m)$ is monotonic and has only one root. Then the optimal value $B_{uv\,opt}$ of parameter B can be found from (2.58) or (2.94). The other parameters of the line are determined by (2.23), (2.27) and (2.45).

Simulation example

In order to verify the proposed switching line design method we consider second order system (2.85) with the previously specified assumptions (2.86), (2.87), (2.88) and $\gamma = 1$. The maximum admissible value of the input signal u_{max} is equal to 5. It means that $U = 1.5$. Furthermore, we require that $v_{max} = 0.5$. For those assumptions we get that $J^v_{IAE}(2) \approx 0.564 > J^u_{IAE}(2) \approx 0.498$ and the optimal parameter m belongs to the interval $(2, m_\gamma)$, where $m_\gamma \approx 3.426$. Finding numerically the root of function $f_1(m)$ given by (2.119) we obtain $m_{uv\,opt} \approx 3.13$. The other optimal switching line parameters are $A_{uv\,opt} \approx 1.64$, $B_{uv\,opt} = -1.5$ and $c_{uv\,opt} \approx 2.87$. The line stops moving at the time instant $t_{f\,uv\,opt}$ equal to 1.09. Simulation results for the system with this line are shown in figures 2.25 – 2.27. It can be seen from figures 2.25 and 2.26 that both constraints are satisfied and that the system is insensitive with respect to the external disturbance and the model uncertainty from the very beginning of the control process. Moreover, the tracking error converges to zero monotonically, without overshoots or oscillations. This property can also be seen from the phase trajectory shown in figure 2.27.

2.2 Switching Line Design Minimising IAE

Fig. 2.25 Tracking error and its derivative

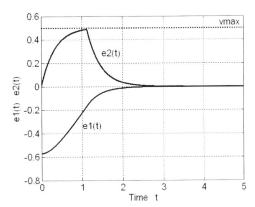

Fig. 2.26 Control signal

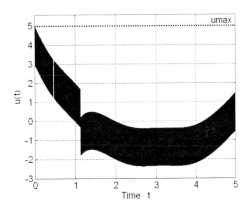

Fig. 2.27 Phase trajectory

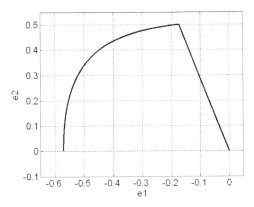

2.3 Switching Line Design Minimising ITAE

In this section, again, switching line parameters are selected. However, now another criterion is taken into account, i.e. the integral of the time multiplied by the absolute error (ITAE) is minimised subject to the input signal and velocity constraints. First, as we did in the previous section, we consider each of the two constraints separately and then we take into account the case when the two constraints must hold simultaneously. We begin our analysis with determining the ITAE

$$J_{ITAE} = \int_{t_0}^{\infty} t |e_1(t)| dt. \tag{2.120}$$

Because the tracking error converges monotonically in the considered system, this criterion can be expressed as

$$J_{ITAE} = \left| \int_0^{\infty} t e_1(t) dt \right|. \tag{2.121}$$

Then substituting equations (2.24) and (2.38) into (2.121) we get

$$
J_{ITAE} = \left| \int_0^{t_f} t e_1(t) dt + \int_{t_f}^{\infty} t e_1(t) dt \right| =
$$

$$
= \left| \int_0^{\frac{e_0 c}{B}} \left(-\frac{B}{c^2} t \, e^{-ct} + e_0 t + \frac{B}{c^2} t - \frac{B}{c} t^2 \right) dt + \int_{\frac{e_0 c}{B}}^{\infty} \left[\frac{B}{c^2} \left(-1 + e^m \right) e^{-ct} t \right] dt \right|. \tag{2.122}
$$

Then calculating appropriate integrals

$$
J_{ITAE} = \left| \left[\frac{B}{c^2} \left(\frac{t}{c} + \frac{1}{c^2} \right) e^{-ct} + \frac{e_0}{2} t^2 + \frac{B}{2 c^2} t^2 - \frac{B}{3 c} t^3 \right]_0^{\frac{e_0 c}{B}} + \right.
$$

$$
\left. + \left[-\frac{B}{c^2} \left(-1 + e^m \right) \left(\frac{t}{c} + \frac{1}{c^2} \right) e^{-ct} \right]_{\frac{e_0 c}{B}}^{\infty} \right|, \tag{2.123}
$$

we obtain

$$
J_{ITAE} = \frac{|e_0|}{6} \left(\frac{6}{c^2} + \frac{3 e_0}{B} + \frac{e_0^2 c^2}{B^2} \right). \tag{2.124}
$$

Finally, substituting (2.45) into (2.124) and taking into account relation (2.28), we get the following form of the considered criterion as a function of variables m and B

2.3 Switching Line Design Minimising ITAE

$$J_{\text{ITAE}}(m, B) = \frac{e_0^2}{6|B|}\left(\frac{6}{m} + 3 + m\right). \tag{2.125}$$

Next, this criterion will be minimised, in order to determine the switching line parameters.

2.3.1 Switching Line Design Subject to Input Signal Constraint

Now, similarly as in section 2.2.1, we consider the situation when u_{max} is the maximum admissible value of the input signal. It means that inequality (2.47) is required to hold, i.e. $\left|u(t)\right| \leq u_{\text{max}}$, where u_{max} is a constant satisfying (2.48). Conditions (2.47) and (2.48) imply that there exists such a strictly positive constant U given by (2.49) that condition (2.47) is satisfied if relation (2.50) holds. Furthermore, relation (2.50) always holds if (2.52) is satisfied. Moreover, from the previous considerations presented in section 2.2.1 we know that $|B|$ is the extreme value of $\left|\dot{e}_2(t)\right|$ and from relation (2.52) we obtain minimisation constraint (2.55). Now we will precisely analyse the minimisation task. Since for any given value of m, the minimum of criterion (2.125) is obtained for the greatest value of $|B|$ satisfying constraint (2.55), therefore the solution of the minimisation task can be found as a minimum of a single variable function $J_{\text{ITAE}}(m)$. Therefore, similarly as in section 2.2.1 we take into account constraint (2.55) and we get the following control quality criterion as a function of a single variable

$$J_{\text{ITAE}}^{u}(m) = \frac{e_0^2}{6\,\delta U}\left(\frac{6}{m} + 3 + m\right). \tag{2.126}$$

Then we calculate the derivative of expression (2.126)

$$\frac{dJ_{\text{ITAE}}^{u}(m)}{dm} = \frac{e_0^2}{6\,\delta U}\left(-\frac{6}{m^2} + 1\right). \tag{2.127}$$

Hence, finally we get the optimal solution $m_{u\,opt} = \sqrt{6}$. At this point criterion (2.126) achieves its minimum value

$$J_{\text{ITAE}}^{u}\left(\sqrt{6}\right) = \frac{e_0^2}{\delta U}\left(\frac{\sqrt{6}}{3} + \frac{1}{2}\right). \tag{2.128}$$

For $m = m_{u\,opt} = \sqrt{6}$, from formula (2.58), we get the optimal value of parameter B which equals $B_{u\,opt} = \delta U \text{sgn}(e_0)$. This concludes the minimisation of criterion (2.125) subject to constraint (2.47). The other optimal parameters of the switching line which can be calculated from relations (2.45) and (2.23), are expressed by the following formulae

$$c_{u\,opt} = \sqrt{\frac{\sqrt{6}\delta U \, \text{sgn}(e_0)}{e_0}} = \sqrt{\frac{\sqrt{6}\delta U}{|e_0|}} \,, \qquad (2.129)$$

$$A_{u\,opt} = -\sqrt{\frac{\sqrt{6}\delta U}{|e_0|}} e_0 = -\text{sgn}(e_0)\sqrt{\sqrt{6}\delta U |e_0|} \,. \qquad (2.130)$$

The optimal switching line arrives at the origin of (e_1, e_2) coordinate frame and becomes fixed at the time instant

$$t_{f\,u\,opt} = \sqrt{\frac{\sqrt{6}|e_0|}{\delta U}} \,. \qquad (2.131)$$

Comparing relations (2.59) and (2.129) one can notice that the switching line slope $c_{u\,opt}$ obtained when minimising ITAE is greater than the same value determined when minimising IAE. Furthermore, taking into account that $B_{u\,opt}$ in both these situations is exactly the same, it may be easily concluded that the switching line designed by minimising ITAE moves horizontally slower towards the origin of the error state space than the line obtained when IAE was considered. Moreover, if the same motion is interpreted as a vertical shift of the line, then in both these cases the line moves with the same velocity $-B_{u\,opt} = -\delta U \text{sgn}(e_0)$ but the line obtained minimising IAE covers a shorter distance than the line optimal in the sense of ITAE. This difference is also reflected in relations (2.61) and (2.131) which directly show that the time $t_{f\,u\,opt}$ calculated in this section is longer than time $t_{f\,u\,opt}$ determined in section 2.2.1. A simulation example presented further in this chapter shows how the difference between the two lines affects the controlled system performance. In general ITAE optimal design results in smaller time constant $1/c$ and reduced forced velocity response of the system $|-B/c|$. Therefore, there is no obvious relation between the quality criterion taken into account in the switching line design process and the system dynamic performance.

Further in the text, similarly as in section 2.2.1, the method of the switching line design, which ensures good dynamic performance of system (2.1) and satisfies elastic input constraint, is presented. In this method we minimise the following criterion which is a sum of the ITAE and some penalty function

$$Q_{ITAE}^u = \int_{t_0}^{\infty} t|e_1(t)|dt + q\left\{\frac{\max[|u(t)|]}{M}\right\}^n \qquad (2.132)$$

where $M > 0$ similarly as in (2.63) is the threshold value of the input signal, $q > 0$ is a weighting factor and $n \geq 1$ is a constant determining how elastic (or stretchable) the constraint is. As it has already been mentioned in section 2.2.1 this criterion reflects both the control quality in the system and the magnitude of control signal $|u(t)|$. Thus, it may be used to design efficient control systems with stretchable input constraints.

2.3 Switching Line Design Minimising ITAE 49

For the purpose of the sliding line design, taking into account relation (2.125), we replace criterion (2.132) with

$$Q_{ITAE}^{u}\left(m,|B|\right) = \frac{e_0^2}{6|B|}\left(\frac{6}{m}+3+m\right)+q\left\{\frac{\max\left[|\dot{e}_2(t)|\right]}{\delta U}\right\}^n.$$
(2.133)

Now we minimise criterion (2.133). Since $\max\left[|\dot{e}_2(t)|\right]$ is expressed by $|B|$, criterion (2.133) can be presented as

$$Q_{ITAE}^{u}\left(m,|B|\right) = \frac{e_0^2}{6|B|}\left(\frac{6}{m}+3+m\right)+q\left(\frac{|B|}{\delta U}\right)^n.$$
(2.134)

This criterion will be minimised by finding its partial derivatives with respect to both variables $(m, |B|)$ and equating these derivatives to zero. The partial derivatives

$$\frac{\partial Q_{ITAE}^{u}\left(m,|B|\right)}{\partial m} = \frac{e_0^2}{6|B|}\left(-\frac{6}{m^2}+1\right)$$
(2.135)

and

$$\frac{\partial Q_{ITAE}^{u}\left(m,|B|\right)}{\partial |B|} = -\frac{e_0^2}{6|B|^2}\left(\frac{6}{m}+3+m\right)+nq\frac{|B|^{n-1}}{\left(\delta U\right)^n}.$$
(2.136)

Solving the following equation

$$\frac{\partial Q_{ITAE}^{u}\left(m,|B|\right)}{\partial m} = 0,$$
(2.137)

we get $m = \sqrt{6}$. Then, from relation

$$\frac{\partial Q_{ITAE}^{u}\left(m,|B|\right)}{\partial |B|} = 0$$
(2.138)

we obtain

$$|B|^{n+1} = \frac{e_0^2\left(\delta U\right)^n}{6nq}\left(\frac{6}{m}+3+m\right).$$
(2.139)

Therefore, the necessary conditions for the existence of the minimum of criterion (2.134) are satisfied at the point $m = \sqrt{6}$ and

$$|B_0| = \left[\frac{e_0^2 (\delta U)^n (2\sqrt{6}+3)}{6qn} \right]^{\frac{1}{n+1}}. \tag{2.140}$$

In order to check if sufficient conditions for the existence of the minimum are also satisfied at this point, we calculate the second order derivatives of criterion Q_{ITAE}^u

$$\frac{\partial^2 Q_{\text{ITAE}}^u \left(m, |B| \right)}{\partial m \partial m} = \frac{2\, e_0^2}{|B|\, m^3}, \tag{2.141}$$

$$\frac{\partial^2 Q_{\text{ITAE}}^u \left(m, |B| \right)}{\partial m \partial |B|} = \frac{\partial^2 Q_{\text{ITAE}}^u \left(m, |B| \right)}{\partial |B| \partial m} = -\frac{e_0^2}{6|B|^2} \left(-\frac{6}{m^2} + 1 \right), \tag{2.142}$$

$$\frac{\partial^2 Q_{\text{ITAE}}^u \left(m, |B| \right)}{\partial |B| \partial |B|} = \frac{e_0^2}{3|B|^3} \left(\frac{6}{m} + 3 + m \right) + qn(n-1)\frac{|B|^{n-2}}{(\delta U)^n}. \tag{2.143}$$

Then, we construct the matrix of the second order derivatives

$$H = \begin{bmatrix} \left. \dfrac{\partial^2 Q_{\text{ITAE}}^u \left(m, |B| \right)}{\partial m \partial m} \right|_{\substack{m=\sqrt{6} \\ |B|=|B_0|}} & \left. \dfrac{\partial^2 Q_{\text{ITAE}}^u \left(m, |B| \right)}{\partial m \partial |B|} \right|_{\substack{m=\sqrt{6} \\ |B|=|B_0|}} \\[4mm] \left. \dfrac{\partial^2 Q_{\text{ITAE}}^u \left(m, |B| \right)}{\partial |B| \partial m} \right|_{\substack{m=\sqrt{6} \\ |B|=|B_0|}} & \left. \dfrac{\partial^2 Q_{\text{ITAE}}^u \left(m, |B| \right)}{\partial |B| \partial |B|} \right|_{\substack{m=\sqrt{6} \\ |B|=|B_0|}} \end{bmatrix}. \tag{2.144}$$

Substituting relations (2.141), (2.142) and (2.143) into (2.144), and putting $m = \sqrt{6}$, $|B| = |B_0|$ we obtain matrix

$$H = \begin{bmatrix} \dfrac{\left(e_0^2 \right)^{\frac{n}{n+1}}}{3\sqrt{6}} \left[\dfrac{6\, qn}{\left(3 + 2\sqrt{6} \right)(\delta U)^n} \right]^{\frac{1}{n+1}} & 0 \\[8mm] 0 & \left[e_0^2 \left(3 + 2\sqrt{6} \right) \right]^{\frac{n-2}{n+1}} \left[\dfrac{6\, qn}{(\delta U)^n} \right]^{\frac{3}{n+1}} \left(\dfrac{1}{3} + \dfrac{n-1}{6} \right) \end{bmatrix} \tag{2.145}$$

Since q, δ, U are positive and n is always greater than or equal to 1, the following two inequalities hold

2.3 Switching Line Design Minimising ITAE

$$\frac{\left(e_0^2\right)^{\frac{n}{n+1}}}{3\sqrt{6}}\left[\frac{6\,qn}{\left(3+2\sqrt{6}\right)\left(\delta U\right)^n}\right]^{\frac{1}{n+1}}>0\,, \tag{2.146}$$

$$\left[e_0^2\left(3+2\sqrt{6}\right)\right]^{\frac{n-2}{n+1}}\left[\frac{6\,qn}{\left(\delta U\right)^n}\right]^{\frac{3}{n+1}}\left(\frac{1}{3}+\frac{n-1}{6}\right)>0\,. \tag{2.147}$$

This implies that matrix H given by (2.144) is positive definite. Hence, for $m=\sqrt{6}$ and

$$B_0=\left[\frac{e_0^2\left(\delta U\right)^n\left(2\sqrt{6}+3\right)}{6qn}\right]^{\frac{1}{n+1}}\mathrm{sgn}\left(e_0\right) \tag{2.148}$$

criterion (2.133) reaches its minimum. This leads to the conclusion that coefficients $m=\sqrt{6}$ and B_0 described by (2.148) are the optimal switching line parameters. The other parameters of the line can be expressed as

$$c_{u\,opt}=\left[\frac{e_0^2\left(\delta U\right)^n\left(2\sqrt{6}+3\right)}{6qn}\right]^{\frac{1}{2n+2}}\sqrt{\frac{\sqrt{6}}{|e_0|}}\,, \tag{2.149}$$

$$A_{u\,opt}=-e_0\left[\frac{e_0^2\left(\delta U\right)^n\left(2\sqrt{6}+3\right)}{6qn}\right]^{\frac{1}{2n+2}}\sqrt{\frac{\sqrt{6}}{|e_0|}}\,. \tag{2.150}$$

The line stops moving and becomes fixed at the time

$$t_{f\,u\,opt}=\sqrt{\sqrt{6}\,|e_0|}\left[\frac{6qn}{e_0^2\left(\delta U\right)^n\left(2\sqrt{6}+3\right)}\right]^{\frac{1}{2n+2}}\,. \tag{2.151}$$

As expected putting $n\to\infty$ into above equations and calculating appropriate limits of (2.148), (2.149), (2.150) and (2.151) we get the switching line parameters expressed by (2.58), (2.129), (2.130) and (2.131), respectively. These parameters ensure the optimal, in the sense of ITAE, performance of system (2.1) subject to the conventional (i.e. non-elastic) input constraint $|u(t)|\le u_{max}$. This conclusion shows again that the conventional constraint considered at the beginning of this section is a special case of the more general elastic constraint analysed here. Criterion $Q_{ITAE}^u\left(m,|B|\right)$ is shown in figure 2.28.

Fig. 2.28 Criterion $Q^u_{ITAE}(m,|B|)$

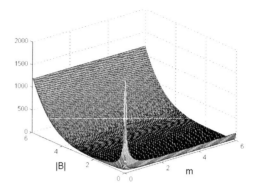

Simulation example

In order to verify how the method proposed in this section works, we consider again system (2.85), with external disturbance (2.86), γ equal to 1, initial conditions (2.87) and the demand trajectory defined by (2.88). The maximum admissible value of the input signal u_{max} is equal to 5, exactly as in the example presented in section 2.2.1. Then $U = 1.5$. The optimal switching line parameters calculated according to formulae (2.58), (2.129) and (2.130) are given as follows $A_{u\,opt} \approx 1.45$, $B_{u\,opt} = -1.5$, $c_{u\,opt} \approx 2.54$ and the line stops moving at the time instant $t_{fu\,opt}$ equal 0.97. Figure 2.29 shows the tracking error and its derivative. It can be seen from the figure that similarly as in the example given in section 2.2.1, here also the error converges to zero monotonically and the system is insensitive with respect to the external disturbance and the model uncertainty from the time $t = 0$. Figure 2.30 presents the input signal which, as required, never exceeds its maximum admissible value $u_{max} = 5$. Figures 2.31 and 2.32 illustrate the system behaviour when elastic input signal constraint with $n = 5$ and $q = 5$ is taken into account. In this case the following switching line parameters $A_{u\,opt} \approx 0.998$, $B_{u\,opt} \approx -0.712$, $c_{u\,opt} \approx 1.748$ and $t_{fu\,opt} \approx 1.402$ are determined. Furthermore, figures 2.33 and 2.34 present the system performance obtained when elastic input signal constraint with $n = 5$ and $q = 0.02$ is considered in the controller design process. In this situation the following parameters of the switching line $A_{u\,opt} \approx 1.581$, $B_{u\,opt} \approx -1.787$, $c_{u\,opt} \approx 2.769$ and $t_{fu\,opt} \approx 0.885$ are obtained. Then, figure 2.35 shows phase trajectories of the system with the same three input constraints, i.e. with the conventional constraint (plot 1) and elastic constraints with $n = 5$, $q = 5$ (plot 2) and $n = 5$, $q = 0.02$ (plot 3). Finally, figures 2.36, 2.37 and 2.38 present a comparison of IAE and ITAE optimal performance of system (2.1) subject to both conventional and elastic input signal constraints. Figure 2.36 shows phase trajectories of the system obtained for conventional input constraint expressed by inequality (2.47) and figures 2.37 and 2.38 show the same trajectories but obtained for elastic input constraint with $n = 5$, $q = 5$ and $n = 5$, $q = 0.02$, respectively.

2.3 Switching Line Design Minimising ITAE

Fig. 2.29 Tracking error and its derivative (conventional input constraint)

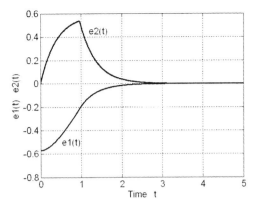

Fig. 2.30 Control signal (conventional input constraint)

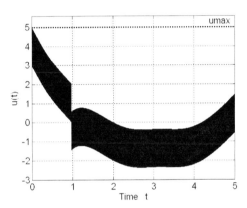

Fig. 2.31 Tracking error and its derivative ($n = 5$, $q = 5$)

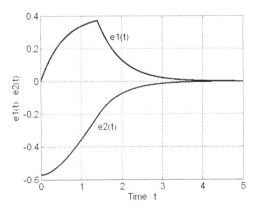

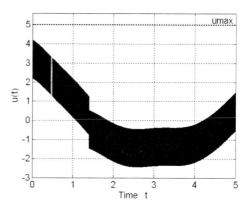

Fig. 2.32 Control signal ($n = 5$, $q = 5$)

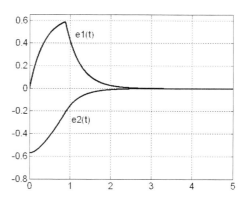

Fig. 2.33 Tracking error and its derivative ($n = 5$, $q = 0.02$)

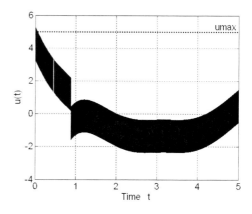

Fig. 2.34 Control signal ($n = 5$, $q = 0.02$)

2.3 Switching Line Design Minimising ITAE

Fig. 2.35 Phase trajectories

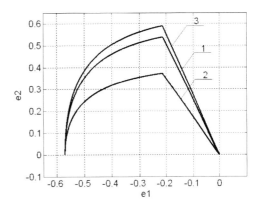

Fig. 2.36 Phase trajectories (conventional input constraint)

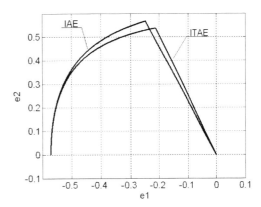

Fig. 2.37 Phase trajectories ($n = 5$, $q = 5$)

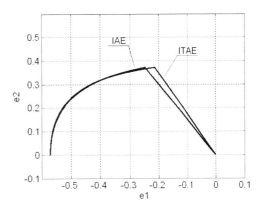

Fig. 2.38 Phase trajectories ($n = 5$, $q = 0.02$)

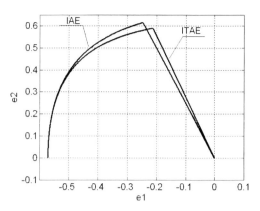

2.3.2 Switching Line Design Subject to Velocity Constraint

In this section we again consider system (2.1) and similarly as in section 2.2.2 we take into account velocity constraint (2.89), i.e. $|e_2(t)| \leq v_{max}$, where v_{max} is the maximum admissible velocity of the system. As we have demonstrated at the beginning of section 2.2.2 this constraint may be expressed by inequality (2.91), which in turn implies that $|B|$ cannot exceed the upper limit given by (2.92). Consequently, substituting (2.92) into (2.125) we obtain

$$J_{ITAE}^v(m) = \frac{|e_0|^3}{6v_{max}^2}(1-e^{-m})^2\left(\frac{6}{m^2}+\frac{3}{m}+1\right). \tag{2.152}$$

This function, specifies the lower bound of criterion $J_{ITAE}(m, B)$ which can be achieved when velocity constraint (2.89) is satisfied.

Since $J_{ITAE}^v(m)$, as shown in figure 2.39, is a decreasing function of its argument, we conclude that the optimal – in the ITAE sense with velocity constraint – value of m tends to infinity, i.e. $m_{v\,opt} \to \infty$. Since parameter B can be calculated

Fig. 2.39 Criterion $J_{ITAE}^v(m)$

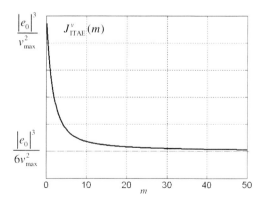

2.3 Switching Line Design Minimising ITAE

substituting $m_{v\,opt}$ into (2.94) then $B_{v\,opt} \to \text{sgn}(e_0)\cdot\infty$, and the other switching line parameters are expressed by (2.96), (2.97) and (2.98). This conclusion shows that when velocity constraint is taken into account, minimisation of ITAE gives exactly the same result as minimisation of IAE. Both of these design methods lead to a vertical switching line which moves horizontally (i.e. along e_1 axis in the error state space) at the rate $-v_{max}\text{sgn}(e_0)$.

Now, similarly as we did in previous subsections, we will modify the considered criterion in order to introduce an elastic velocity constraint. This modified criterion may be expressed as

$$Q^v_{ITAE}\left(m,|B|\right) = \int_{t_0}^{\infty} t\left|e_1\left(t\right)\right|dt + q\left\{\frac{\max\left[\left|e_2\left(t\right)\right|\right]}{M}\right\}^n \tag{2.153}$$

where M, q and n are defined exactly as in section 2.2.2, i.e. $M > 0$ is the threshold value of the system velocity, $q > 0$ is a weighting factor and $n \geq 1$ is a constant determining how elastic the constraint is. Since v_{max} is the threshold value of the system velocity, we obtain

$$Q^v_{ITAE}\left(m,|B|\right) = \frac{e_0^2}{6|B|}\left(\frac{6}{m}+3+m\right) + q\left\{\frac{\max\left[\left|e_2\left(t\right)\right|\right]}{v_{max}}\right\}^n. \tag{2.154}$$

Now we minimise criterion (2.154). Taking into account (2.45) and (2.90), we get

$$Q^v_{ITAE}\left(m,|B|\right) = \frac{e_0^2}{6|B|}\left(\frac{6}{m}+3+m\right) + q\left[\frac{\sqrt{|B||e_0|}\left(1-e^{-m}\right)}{v_{max}\sqrt{m}}\right]^n. \tag{2.155}$$

Our further analysis will be quite similar to the reasoning presented in section 2.2.2. First we notice that for any m, derivative

$$\frac{\partial Q^v_{ITAE}\left(m,|B|\right)}{\partial|B|} = -\frac{e_0^2}{6B^2}\left(\frac{6}{m}+3+m\right) + \frac{1}{2}qn|B|^{n/2-1}\left[\frac{\sqrt{|e_0|}\left(1-e^{-m}\right)}{v_{max}\sqrt{m}}\right]^n \tag{2.156}$$

is negative for small values of $|B|$, and for all values of this argument which are bigger than the only root of equation

$$\frac{\partial Q^v_{ITAE}\left(m,|B|\right)}{\partial|B|} = 0 \tag{2.157}$$

the derivative is positive. Therefore, for any m, criterion Q^v_{ITAE} reaches its minimum for

$$|B| = \left[\frac{e_0^2 \left(\dfrac{6}{m} + 3 + m \right) \left(v_{max} \sqrt{m} \right)^n}{3\,qn \left(\sqrt{|e_0|} \right)^n \left(1 - e^{-m} \right)^n} \right]^{\frac{2}{n+2}}.$$

(2.158)

Then substituting (2.158) into criterion (2.155) we get

$$Q_{ITAE}^v \left(m, |B| \right) = \frac{e_0^2 \left[3\,qn \left(\sqrt{|e_0|} \right)^n \left(1 - e^{-m} \right)^n \right]^{\frac{2}{n+2}}}{6 \left[e_0^2 \left(\dfrac{6}{m} + 3 + m \right) \left(v_{max} \sqrt{m} \right)^n \right]^{\frac{2}{n+2}}} \left(\frac{6}{m} + 3 + m \right) +$$

$$+ q \left[\frac{\sqrt{|e_0|} \left(1 - e^{-m} \right)^2}{v_{max} \sqrt{m}} \right]^n \left[\frac{e_0^2 \left(\dfrac{6}{m} + 3 + m \right) \left(v_{max} \sqrt{m} \right)^n}{3\,qn \left(\sqrt{|e_0|} \right)^n \left(1 - e^{-m} \right)^n} \right]^{\frac{n}{n+2}} =$$

$$= \left[\frac{|e_0|^3 \left(1 - e^{-m} \right)^2}{3\,v_{max}^2\, m} \left(\frac{6}{m} + 3 + m \right) \right]^{\frac{n}{n+2}} (nq)^{\frac{2}{n+2}} \left(\frac{1}{2} + \frac{1}{n} \right) =$$

(2.159)

$$= \left[\frac{|e_0|^3 \left(1 - e^{-m} \right)^2}{3\,v_{max}^2} \left(\frac{6}{m^2} + \frac{3}{m} + 1 \right) \right]^{\frac{n}{n+2}} (nq)^{\frac{2}{n+2}} \left(\frac{1}{2} + \frac{1}{n} \right).$$

Since the expression in the square bracket equals one half of J_{ITAE}^v given by (2.152) which is a decreasing function of m, we conclude that function (2.154) achieves its minimum for $m_{v\,opt} \to \infty$. Then substituting $m_{v\,opt} \to \infty$ into relation (2.158) we find that the optimal values of parameters $|B|$ and c also tend to infinity and $A_{v\,opt} \to -\,sgn(e_0) \cdot \infty$.

Consequently, we get

$$t_{f\,v\,opt} = \left(\frac{3\,qn |e_0|^{n-1}}{v_{max}^n} \right)^{\frac{1}{n+2}}.$$

(2.160)

When $n \to \infty$ then $t_{fv\,opt}$ tends to expression (2.98) which specifies the time of the line motion both in the case of ITAE minimisation and IAE minimisation with conventional velocity constraint. This conclusion ends the task of ITAE

2.3 Switching Line Design Minimising ITAE

Fig. 2.40 Criterion $Q_{ITAE}^v(m,|B|)$

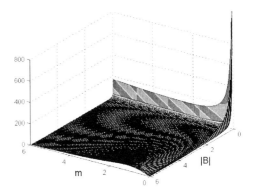

minimisation with the elastic velocity constraint. Criterion Q_{ITAE}^v is illustrated in figure 2.40.

Simulation example

In this example again we consider system (2.85), with external disturbance given by (2.86), $\gamma = 1$, the initial conditions given by (2.87) and the demand trajectory defined by (2.88). The maximum admissible value of the system velocity v_{max} is equal to 0.5. Taking into account the conventional velocity constraint expressed by an inequality, we obtain the same parameters of the switching line $A_{v\,opt} \to \infty$, $B_{v\,opt} \to -\infty$, $c_{v\,opt} \to \infty$ and $t_{fv\,opt} \approx 1.142$ as previously in section 2.2.2. Consequently, the system behaviour is illustrated in figures 2.12 – 2.15. However, when elastic velocity constraint is taken into account the switching line optimal in the sense of ITAE moves at a different pace than, determined in the example presented in section 2.2.2, switching line optimal in the sense of IAE. Indeed, if $n = 5$, $q = 5$, then $t_{fv\,opt} \approx 2.772$ and for $n = 5$, $q = 0.02$ the time $t_{fv\,opt} \approx 1.26$. The system behaviour for $n = 5$, $q = 5$ is shown in figures 2.41 – 2.44 and figures 2.45 – 2.48 present the performance of the same system when $n = 5$, $q = 0.02$.

Fig. 2.41 Tracking error ($n = 5$, $q = 5$)

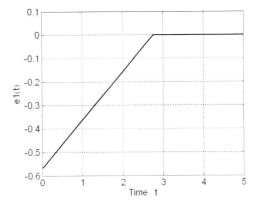

Fig. 2.42 Tracking error derivative ($n = 5$, $q = 5$)

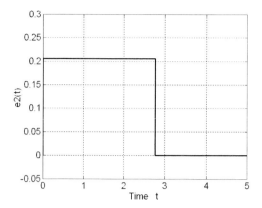

Fig. 2.43 Phase trajectory ($n = 5$, $q = 5$)

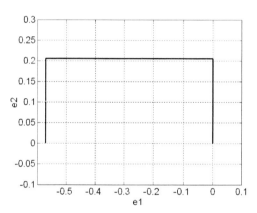

Fig. 2.44 Control signal ($n = 5$, $q = 5$)

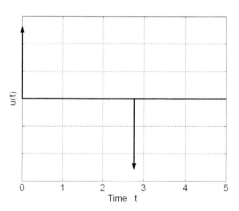

2.3 Switching Line Design Minimising ITAE

Fig. 2.45 Tracking error ($n = 5$, $q = 0.02$)

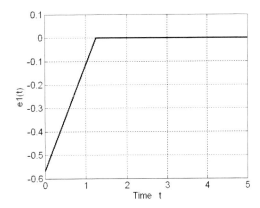

Fig. 2.46 Tracking error derivative ($n = 5$, $q = 0.02$)

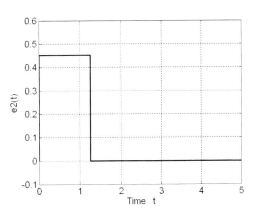

Fig. 2.47 Phase trajectory ($n = 5$, $q = 0.02$)

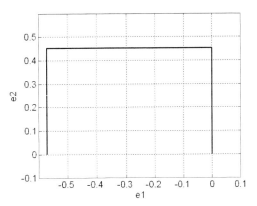

Fig. 2.48 Control signal ($n = 5$, $q = 0.02$)

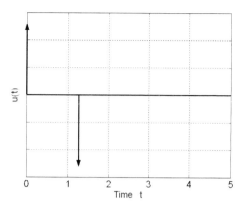

In this way we determined the optimal switching line parameters minimising ITAE with each of the constraints taken into account separately. In the next section we consider the case when both the constraints are required to be satisfied simultaneously.

2.3.3 Switching Line Design Subject to Input Signal and Velocity Constraints

In this section we design the switching line optimal in the sense of ITAE and we take into account both input signal and velocity constraints. In other words, we minimise ITAE when u_{max} is the maximum admissible value of the input signal, and the velocity is required not to exceed v_{max}, i.e. we take into account both constraints (2.47) and (2.89). In order to find the minimum value of criterion $J_{ITAE}(m, B)$ expressed by (2.125) we will minimise the maximum of $J_{ITAE}^u(m)$ and $J_{ITAE}^v(m)$, i.e. we will find the minimum of the following function

$$J_{ITAE}^{uv} = \max\left[J_{ITAE}^u(m), J_{ITAE}^v(m)\right] \tag{2.161}$$

where $J_{ITAE}^u(m)$ and $J_{ITAE}^v(m)$ are given by (2.126) and (2.152), respectively. The optimal solution of the minimisation problem considered in this section will be such a value $m_{uv\,opt}$ that

$$J_{ITAE}^{uv}(m_{uv\,opt}) = \min_{m>0}\left\{\max\left[J_{ITAE}^u(m), J_{ITAE}^v(m)\right]\right\} \tag{2.162}$$

and the respective value of parameter B. This shows that the switching line design procedure adopted in this section is quite similar to the one used in section 2.2.3. We consider the following two cases: $J_{ITAE}^v(\sqrt{6}) \leq J_{ITAE}^u(\sqrt{6})$ and $J_{ITAE}^v(\sqrt{6}) > J_{ITAE}^u(\sqrt{6})$. In the case when $J_{ITAE}^v(\sqrt{6}) \leq J_{ITAE}^u(\sqrt{6})$ we get

2.3 Switching Line Design Minimising ITAE

$$J_{ITAE}^{uv}\left(m_{uv\,opt}\right) = \min_{m>0}\left\{\max\left[J_{ITAE}^{u}(m), J_{ITAE}^{v}(m)\right]\right\} = J_{ITAE}^{u}\left(\sqrt{6}\right). \quad (2.163)$$

Hence the optimal value of parameter m is $m_{uv\,opt} = m_{u\,opt} = \sqrt{6}$ and the optimal value $B_{uv\,opt} = B_{u\,opt}$ can be found from (2.58).

In the latter case when $J_{ITAE}^{v}\left(\sqrt{6}\right) > J_{ITAE}^{u}\left(\sqrt{6}\right)$, we notice that function $J_{ITAE}^{u}(m)$ increases for any m greater than $\sqrt{6}$, and

$$\lim_{m \to \infty} J_{ITAE}^{u}(m) = \infty. \quad (2.164)$$

Furthermore, function $J_{ITAE}^{v}(m)$ decreases for every value of m and

$$\lim_{m \to \infty} J_{ITAE}^{v}(m) = \frac{|e_0|^3}{6v_{max}^2}. \quad (2.165)$$

This implies that the minimum $J_{ITAE}^{uv}(m)$ is achieved at a point $m_{uv\,opt}$, such that $m_{uv\,opt} > \sqrt{6}$ and the following condition holds $J_{ITAE}^{u}\left(m_{uv\,opt}\right) = J_{ITAE}^{v}\left(m_{uv\,opt}\right)$. The reasoning presented here is illustrated in figure 2.49.

Now we will show how to find the optimal value $m_{uv\,opt}$. For any $m > 0$, criterion $J_{ITAE}^{v}(m)$ is upper bounded by the following inequality

$$J_{ITAE}^{v}(m) = \frac{|e_0|^3}{6v_{max}^2}\left(\frac{6}{m^2} + \frac{3}{m} + 1\right)(1 - e^{-m})^2 < \frac{|e_0|^3}{6v_{max}^2}\left(\frac{6}{m^2} + \frac{3}{m} + 1\right). \quad (2.166)$$

Thus, solving equation

$$\frac{|e_0|^3}{6v_{max}^2}\left(\frac{6}{m^2} + \frac{3}{m} + 1\right) = J_{ITAE}^{u}(m) \quad (2.167)$$

Fig. 2.49 Criteria $J_{ITAE}^{v}(m)$ and $J_{ITAE}^{u}(m)$

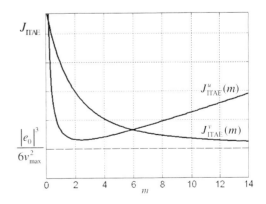

which is equivalent to

$$\frac{\left|e_0\right|^3}{6v_{max}^2}\frac{1}{m}\left(\frac{6}{m}+3+m\right)=\frac{e_0^2}{6\delta U}\left(\frac{6}{m}+3+m\right),\qquad(2.168)$$

we obtain such a value of m that $m_{uv\,opt}$ is greater than $\sqrt{6}$ and smaller than this value. Actually this value is given by (2.118) and consequently $m_{uv\,opt}\in(\sqrt{6}\,,m_\gamma)$.

In order to find $m_{uv\,opt}$ the root of the following function

$$f_2(m)=J_{ITAE}^v(m)-J_{ITAE}^u(m)=\frac{e_0^2}{6}\left(\frac{6}{m}+3+m\right)\left[\frac{\left|e_0\right|\left(1-e^{-m}\right)^2}{v_{max}^2\,m}-\frac{1}{\delta U}\right]\qquad(2.169)$$

is determined numerically in the interval ($\sqrt{6}\,,m_\gamma$). In fact, in this interval $f_2(m)$ is monotonic and has only one root. Then the optimal value $B_{uv\,opt}$ of parameter B can be found from (2.58) or (2.94) while $c_{uv\,opt}$, $A_{uv\,opt}$ and $t_{f\,uv\,opt}$ may be determined from (2.45), (2.23) and (2.27), respectively.

Notice that when $J_{ITAE}^v\left(\sqrt{6}\right)\geq J_{ITAE}^u\left(\sqrt{6}\right)$, using the presented method we get the same switching line parameters as in section 2.2.3. This is a direct consequence of the fact that the roots of functions $f_1(m)$ and $f_2(m)$ for any fixed values of e_0, v_{max}, δ and U are exactly the same, which can be seen from the following relation

$$\frac{\left|e_0\right|\left(1-e^{-m}\right)^2}{v_{max}^2\,m}-\frac{1}{\delta U}=\frac{1}{m}\left[\frac{\left|e_0\right|\left(1-e^{-m}\right)^2}{v_{max}^2}-\frac{m}{\delta U}\right]=$$
$$=\frac{1}{m}\left[\frac{\sqrt{\left|e_0\right|}\left(1-e^{-m}\right)}{v_{max}}-\sqrt{\frac{m}{\delta U}}\right]\left[\frac{\sqrt{\left|e_0\right|}\left(1-e^{-m}\right)}{v_{max}}+\sqrt{\frac{m}{\delta U}}\right].$$
$$(2.170)$$

Therefore, the system performance when the switching line is selected according to the procedure presented in this section is illustrated by simulation example given in section 2.2.3.

It is worth to point out that only condition $J_{ITAE}^v\left(\sqrt{6}\right)\geq J_{ITAE}^u\left(\sqrt{6}\right)$ ensures that IAE and ITAE optimal solutions are identical, while condition $J_{ITAE}^v(2)\geq J_{ITAE}^u(2)$ does not imply that the two solutions are the same. The relation between the IAE and ITAE optimal solutions is shown in table 2.1, where parameter $m_\gamma=\left|e_0\right|\delta U/v_{max}^2$ is defined by (2.118) and its characteristic values are given as follows

$$m_{\gamma1}=2\left(1-e^{-2}\right)^{-2}\approx2.675,\qquad(2.171)$$

$$m_{\gamma2}=\sqrt{6}\left(1-e^{-\sqrt{6}}\right)^{-2}\approx2.934.\qquad(2.172)$$

2.3 Switching Line Design Minimising ITAE

Table 2.1 Relation between IAE and ITAE optimal solutions

m_γ	$m_\gamma \in \left(0, m_{\gamma 1}\right]$	$m_\gamma \in \left(m_{\gamma 1}, m_{\gamma 2}\right]$	$m_\gamma \in \left(m_{\gamma 2}, \infty\right)$
IAE optimal $m_{uv\,opt}$	2	determined numerically $m_{uv\,opt} \in \left(2, \sqrt{6}\,\right]$	determined numerically $m_{uv\,opt} \in \left(\sqrt{6}, m_\gamma\right)$
ITAE optimal $m_{uv\,opt}$	$\sqrt{6}$	$\sqrt{6}$	
J_{IAE}^{v} vs. J_{IAE}^{u}	$J_{\mathrm{IAE}}^{v}\left(2\right) \leq J_{\mathrm{IAE}}^{u}\left(2\right)$	$J_{\mathrm{IAE}}^{v}\left(2\right) > J_{\mathrm{IAE}}^{u}\left(2\right)$	$J_{\mathrm{IAE}}^{v}\left(2\right) > J_{\mathrm{IAE}}^{u}\left(2\right)$
J_{ITAE}^{v} vs. J_{ITAE}^{u}	$J_{\mathrm{ITAE}}^{v}\left(\sqrt{6}\right) < \ < J_{\mathrm{ITAE}}^{u}\left(\sqrt{6}\right)$	$J_{\mathrm{ITAE}}^{v}\left(\sqrt{6}\right) \leq \ \leq J_{\mathrm{ITAE}}^{u}\left(\sqrt{6}\right)$	$J_{\mathrm{ITAE}}^{v}\left(\sqrt{6}\right) > \ > J_{\mathrm{ITAE}}^{u}\left(\sqrt{6}\right)$

Chapter 3
Time-Varying Sliding Modes for the Third Order Systems

In this chapter we further develop the sliding mode design method for possibly time-varying and nonlinear continuous time plants. We still consider the linear sliding surfaces, however now we focus our attention on more complex, i.e. third order systems, described by the following equations

$$\dot{x}_1 = x_2$$
$$\dot{x}_2 = x_3 \quad (3.1)$$
$$\dot{x}_3 = f(x,t) + \Delta f(x,t) + b(x,t)u + d(t)$$

where x_1, x_2, x_3 are the state variables of the system and $x(t) = [x_1(t)\ x_2(t)\ x_3(t)]^T$ is the state vector. Similarly as in chapter 2, also in this part of the book t denotes time, u is the input signal, b, f – are a priori known, bounded functions of time and the system state, Δf and d are functions representing the system uncertainty and external disturbances, respectively. Not only the same notation as in chapter 2 will be used further in this book, but also the same assumptions that there exists a strictly positive constant δ which is the lower bound of $b(x, t)$, and that functions Δf and d are unknown and bounded by a constant μ are made. In other words we assume that relations (2.3) and (2.4) hold for every pair (x, t). Initial conditions of the system are

$$x_{10} = x_1(t_0),\ x_{20} = x_2(t_0),\ x_{30} = x_3(t_0). \quad (3.2)$$

The system (3.1) should follow the desired trajectory

$$x_d(t) = [x_{1d}(t)\ x_{2d}(t)\ x_{3d}(t)]^T \quad (3.3)$$

where $x_{2d}(t) = \dot{x}_{1d}(t)$, $x_{3d}(t) = \dot{x}_{2d}(t)$ and $x_{3d}(t)$ is a differentiable function of time. The trajectory tracking error may be expressed as

$$e(t) = [e_1(t)\ e_2(t)\ e_3(t)]^T = x(t) - x_d(t). \quad (3.4)$$

Consequently,

$$e_1(t) = x_1(t) - x_{1d}(t),\ e_2(t) = x_2(t) - x_{2d}(t),\ e_3(t) = x_3(t) - x_{3d}(t). \quad (3.5)$$

In this chapter it is assumed that the initial tracking error and its derivatives satisfy the following equations

$$e_1(t_0) = e_0,\ e_2(t_0) = 0,\ e_3(t_0) = 0 \quad (3.6)$$

A. Bartoszewicz and A. Nowacka-Leverton: Time-Varying Sliding Modes, LNCIS 382, pp. 67–179.
springerlink.com
© Springer-Verlag Berlin Heidelberg 2009

68   3 Time-Varying Sliding Modes for the Third Order Systems

where, similarly as it was assumed in the previous chapter, e_0 is an arbitrary real number different from zero. Since in this chapter we consider the third order system, we can no longer take into account switching lines, but in the next section we introduce a time-varying switching plane.

3.1 Control Strategy

Let us consider a time-varying switching plane with the constant angle of inclination. Originally this plane moves with a constant velocity in the state space and then it stops at the time t_f and remains fixed. Consequently, for any $t \le t_f$ the switching plane is defined as

$$s(e,t) = 0 \text{ where } s(e,t) = e_3(t) + c_2 e_2(t) + c_1 e_1(t) + A + Bt \qquad (3.7)$$

where c_1, c_2, A and B are some constants, which will be selected further in the chapter. Since the considered plane stops at the time t_f, for any $t \ge t_f$ it is described as

$$s(e,t) = 0 \text{ where } s(e,t) = e_3(t) + c_2 e_2(t) + c_1 e_1(t). \qquad (3.8)$$

First, similarly as in the previous chapter, the constants c_1, c_2, A and B should be chosen in such a way that the representative point of the system at the initial time $t = t_0$ belongs to the switching plane. Therefore, the following condition should hold

$$s\left[e\left(t_0\right),t_0\right] = e_3\left(t_0\right) + c_2 e_2\left(t_0\right) + c_1 e_1\left(t_0\right) + A + Bt_0 = 0. \qquad (3.9)$$

Similarly as in chapter 2, it can be shown that the input signal

$$u = \frac{-f(x,t) - c_2 e_3(t) - c_1 e_2(t) + \dot{x}_{3d}(t) - B - \gamma \mathrm{sgn}\left[s(e,t)\right]}{b(x,t)} \qquad (3.10)$$

where $\gamma = \eta + \mu$ and η is a strictly positive constant, ensures the stability of the sliding motion on the switching plane, introduced in this section. In order to demonstrate this property we consider the product

$$s(e,t)\dot{s}(e,t) = s(e,t)\left[\dot{e}_3(t) + c_2 e_3(t) + c_1 e_2(t) + B\right]. \qquad (3.11)$$

Substituting relations (3.1), (3.4) and (3.10) into (3.11) we obtain the following inequality

$$s(e,t)\dot{s}(e,t) = s(e,t)\left\{\Delta f(x,t) + d(t) - \gamma \mathrm{sgn}\left[s(e,t)\right]\right\} \le -\eta\left|s(e,t)\right| \qquad (3.12)$$

which proves the existence and stability of the sliding motion on the plane described by equations (3.7) and (3.8). Consequently, we may conclude that for any

3.1 Control Strategy

time $t \in \langle 0, t_f \rangle$ the system dynamics is described by equation (3.7) with initial conditions (3.6). Thus, we take into account the following equation

$$e_3(t) + c_2 e_2(t) + c_1 e_1(t) + A + Bt = 0. \tag{3.13}$$

In order to solve it, we consider homogenous equation

$$e_3(t) + c_2 e_2(t) + c_1 e_1(t) = 0. \tag{3.14}$$

Since fast tracking error convergence to zero without oscillations is required, the characteristic polynomial of equation (3.14) should have one, double real root. This ensures that oscillations will take place neither before the switching plane stops, nor after it does. Hence, we get another condition

$$c_2 = 2\sqrt{c_1}. \tag{3.15}$$

Furthermore, parameters c_1 and c_2 must be strictly positive to make system (3.1) stable in the sliding mode. Solving equation (3.13) with condition (3.15) and assuming for the sake of clarity that $t_0 = 0$ we obtain the tracking error and its first and second derivatives for the time $t \in \langle 0, t_f \rangle$

$$e_1(t) = \left[e_0 + \frac{A}{c_1} - \frac{2B\sqrt{c_1}}{c_1^2} + \left(\frac{A}{\sqrt{c_1}} + \sqrt{c_1}\, e_0 - \frac{B}{c_1} \right) t \right] e^{-\sqrt{c_1}\, t} + \tag{3.16}$$
$$- \frac{A}{c_1} + \frac{2B\sqrt{c_1}}{c_1^2} - \frac{B}{c_1} t,$$

$$e_2(t) = \left[\frac{B}{c_1} - \left(A + c_1 e_0 - \frac{B\sqrt{c_1}}{c_1} \right) t \right] e^{-\sqrt{c_1}\, t} - \frac{B}{c_1}, \tag{3.17}$$

$$e_3(t) = \left[-A - c_1 e_0 + c_1 \left(\frac{A}{\sqrt{c_1}} + e_0\sqrt{c_1} - \frac{B}{c_1} \right) t \right] e^{-\sqrt{c_1}\, t}. \tag{3.18}$$

Taking into account conditions (3.6), (3.9) and the assumption that $t_0 = 0$ one gets expression

$$A = -c_1 e_0 \tag{3.19}$$

which is directly related to relation (2.23) in chapter 2. Taking into account expression (3.19), formulae (3.16), (3.17) and (3.18) can be expressed as follows

$$e_1(t) = \left(-\frac{2B\sqrt{c_1}}{c_1^2} - \frac{B}{c_1} t \right) e^{-\sqrt{c_1}\, t} + \frac{2B\sqrt{c_1}}{c_1^2} + e_0 - \frac{B}{c_1} t, \tag{3.20}$$

$$e_2(t) = \frac{B}{c_1}\left(1 + \sqrt{c_1}\, t\right) e^{-\sqrt{c_1}\, t} - \frac{B}{c_1}, \tag{3.21}$$

$$e_3(t) = -Bt e^{-\sqrt{c_1}\, t}. \tag{3.22}$$

Having found the tracking error and its derivatives for $t \leq t_f$ we move on to the analysis of the system behaviour in the second phase of its motion, that is when the switching plane does not move. Since for the time $t \geq t_f$ the switching plane is fixed and passes through the origin of the error state space, then condition (2.26) holds in the situation considered in this chapter, i.e. $A + Bt_f = 0$. From equations (2.26) and (3.19) we obtain relation

$$t_f = \frac{e_0 c_1}{B} \tag{3.23}$$

which is quite similar to equation (2.27). Since c_1 and t_f are strictly positive, relation (3.23) leads to the conclusion that e_0 and B have the same signs. The fixed sliding plane is described by relation (3.8) and also by equation (3.14). The initial conditions required to solve equation (3.14) may be found from equations (3.20), (3.21) and (3.22) substituting t_f for independent variable t. With the following notation

$$k = \frac{c_1 \sqrt{c_1}\, e_0}{B}, \tag{3.24}$$

the initial conditions for the second phase of the system motion can be written as

$$e_1(t_f) = \left(-\frac{2B\sqrt{c_1}}{c_1^2} - e_0\right) e^{-k} + \frac{2B\sqrt{c_1}}{c_1^2}, \tag{3.25}$$

$$e_2(t_f) = -\frac{B}{c_1} + \left(\frac{B}{c_1} + e_0 \sqrt{c_1}\right) e^{-k}, \tag{3.26}$$

$$e_3(t_f) = -e^{-k} c_1 e_0. \tag{3.27}$$

The parameter k defined by equation (3.24) is strictly positive. Solving equation (3.14) with initial conditions (3.25), (3.26), (3.27) and using relation (3.19) we get

$$e_1(t) = e^{-\sqrt{c_1}\, t}\left[-\frac{2B\sqrt{c_1}}{c_1^2} + \frac{2B\sqrt{c_1}}{c_1^2} e^k - e_0 e^k + \left(-\frac{B}{c_1} + \frac{B}{c_1} e^k\right) t\right], \tag{3.28}$$

3.1 Control Strategy

$$e_2(t) = e^{-\sqrt{c_1}\,t}\left[\frac{B}{c_1} - \frac{B}{c_1}e^k + e_0 e^k \sqrt{c_1} - \left(-\frac{B}{\sqrt{c_1}} + \frac{B}{\sqrt{c_1}}e^k\right)t\right],\tag{3.29}$$

$$e_3(t) = e^{-\sqrt{c_1}\,t}\left[-e_0 e^k c_1 + B\left(e^k - 1\right)t\right].\tag{3.30}$$

These formulae describe the tracking error and its derivatives for any time $t \geq t_f$. It is important that the error described by equations (3.20) and (3.28) does not exhibit any overshoots. This can be demonstrated as follows. For the time $t \in \langle 0, t_f\rangle$ the tracking error is described by equation (3.20). During this interval the error either monotonically decreases for $e_0 > 0$ to a value $e_1(t_f) > 0$ or monotonically increases for $e_0 < 0$ to a value $e_1(t_f) < 0$. This can be seen from equations (3.21) and (3.25) reformulated as

$$e_2(t) = \frac{B}{c_1}e^{-\sqrt{c_1}\,t}\left[\left(1 + \sqrt{c_1}\,t\right) - e^{\sqrt{c_1}\,t}\right],\tag{3.31}$$

$$e_1(t_f) = \frac{2B\sqrt{c_1}}{c_1^2}e^{-k}\left(-1 - \frac{e_0 c_1 \sqrt{c_1}}{2B} + e^k\right) = \frac{2B\sqrt{c_1}}{c_1^2}e^{-k}\left(-1 - \frac{k}{2} + e^k\right) =$$

$$= \frac{2B\sqrt{c_1}}{c_1^2}e^{-k}\left[-\left(1 + \frac{k}{2}\right) + e^k\right].\tag{3.32}$$

The signs of B and e_0 are the same, so it can be seen from equation (3.31) that the derivative of the tracking error has the opposite sign to e_0 for any $t \in \langle 0, t_f\rangle$. Equation (3.32) shows that the signs of the tracking error at the time instants t_0 and t_f are the same. Hence, the error does not exhibit any overshoot during this time interval. On the other hand, for the time $t \geq t_f$ the tracking error is given by equation (3.28). If $e_0 > 0$, then the tracking error decreases from the positive value $e_1(t_f)$ to zero. If $e_0 < 0$, then the error increases from the negative value $e_1(t_f)$ to zero. The tracking error converges to zero monotonically because from equation (3.29)

$$e_2(t) = \frac{B}{c_1}e^{-\sqrt{c_1}\,t}\left[1 - e^k + ke^k - \sqrt{c_1}\left(e^k - 1\right)t\right],\tag{3.33}$$

and for any $t \geq t_f$, the expression in the square bracket $1 - e^k + ke^k - \sqrt{c_1}\left(e^k - 1\right)t$

$$\leq 1 - e^k + ke^k - \sqrt{c_1}\left(e^k - 1\right)t_f = 1 - e^k + ke^k - \sqrt{c_1}\left(e^k - 1\right)e_0 c_1/B = 1 - e^k + k \quad \text{is}$$

negative. Thus it follows from equation (3.33) that the derivative of the tracking error has the opposite sign to e_0. Next in this chapter the new methods of the switching plane selection will be proposed.

3.2 Switching Plane Design Minimising IAE

In this section the switching plane parameters are selected. This is done in a similar way as in section 2.2 since here the integral of the absolute error (IAE) is minimised again. However, now we take into account three constraints, i.e. we consider the input signal, acceleration and velocity constraints. First we take into account each of the three constraints separately, then we require each pair of these constraints to be satisfied at a time, and finally, we take into account the case when all the three constraints must hold simultaneously. We begin with calculating the IAE. Substituting equations (3.20) and (3.28) into (2.41), we obtain

$$
J_{\mathrm{IAE}} = \left| \int_0^{t_f} e_1(t)\,dt + \int_{t_f}^{\infty} e_1(t)\,dt \right| =
$$

$$
= \left| \int_0^{\frac{e_0 c_1}{B}} \left[\left(-\frac{2B\sqrt{c_1}}{c_1^2} - \frac{B}{c_1} t \right) e^{-\sqrt{c_1}\,t} + \frac{2B\sqrt{c_1}}{c_1^2} + e_0 - \frac{B}{c_1} t \right] dt + \right. \tag{3.34}
$$

$$
\left. + \int_{\frac{e_0 c_1}{B}}^{\infty} \left\{ e^{-\sqrt{c_1}\,t} \left[\frac{2B\sqrt{c_1}}{c_1^2} e^k - \frac{2B\sqrt{c_1}}{c_1^2} - e_0 e^k + \left(-\frac{B}{c_1} + \frac{B}{c_1} e^k \right) t \right] \right\} dt \right|.
$$

Then, after some straightforward calculations, we obtain

$$
J_{\mathrm{IAE}} = \frac{2|e_0|}{\sqrt{c_1}} + \frac{e_0^2 c_1}{2|B|}. \tag{3.35}
$$

Moreover, from equation (3.24) we get

$$
c_1 = \left(\frac{Bk}{e_0} \right)^{2/3}. \tag{3.36}
$$

Therefore, in order to simplify the minimisation process, similarly as we did in chapter 2, also in this chapter we may consider the quality criterion as a function of variables k and B, instead of c_1 and B. Indeed, substituting (3.36) into (3.35), we obtain

$$
J_{\mathrm{IAE}}(k, B) = \frac{|e_0|^{4/3}}{|B|^{1/3}} \left(2k^{-1/3} + \frac{1}{2} k^{2/3} \right). \tag{3.37}
$$

Further in the text, this criterion will be minimised subject to various constraints. In this way the moving switching plane parameters will be determined.

3.2.1 Switching Plane Design Subject to Input Signal Constraint

In this section, similarly as in section 2.2.1 we consider only input signal constraint, i.e. we assume that u_{max} is the maximum admissible value of u. This constraint is expressed by inequality (2.47) where the constant u_{max} is big enough to satisfy the following condition

$$u_{max} > \frac{\left|\dot{x}_{3d} - f(\boldsymbol{x},t)\right| + \gamma}{\left|b(\boldsymbol{x},t)\right|}. \tag{3.38}$$

Inequalities (2.47) and (3.38) ensure that there exists such a constant

$$U = u_{max} - \max\left[\frac{\left|\dot{x}_{3d} - f(\boldsymbol{x},t)\right| + \gamma}{\left|b(\boldsymbol{x},t)\right|}\right] \tag{3.39}$$

strictly greater than zero that the following relation

$$\left|\dot{e}_3(t)\right| \le \left|b(\boldsymbol{x},t)\right| U \tag{3.40}$$

implies that inequality (2.47) is satisfied. Since from relations (3.7), (3.10), (3.39) and (3.40) we have

$$\left|u(t)\right| \le \frac{1}{\left|b(\boldsymbol{x},t)\right|}\left[\left|\dot{e}_3\right| + \left|\dot{x}_{3d} - f(\boldsymbol{x},t)\right| + \gamma\right] \le$$

$$\le \frac{1}{\left|b(\boldsymbol{x},t)\right|}\left\{\left|b(\boldsymbol{x},t)\right| U + \left[\left|\dot{x}_{3d} - f(\boldsymbol{x},t)\right| + \gamma\right]\right\} =$$

$$= U + \frac{1}{\left|b(\boldsymbol{x},t)\right|}\left[\left|\dot{x}_{3d} - f(\boldsymbol{x},t)\right| + \gamma\right] \le U + \max\left[\frac{\left|\dot{x}_{3d} - f(\boldsymbol{x},t)\right| + \gamma}{\left|b(\boldsymbol{x},t)\right|}\right] = \tag{3.41}$$

$$= u_{max} - \max\left[\frac{\left|\dot{x}_{3d} - f(\boldsymbol{x},t)\right| + \gamma}{\left|b(\boldsymbol{x},t)\right|}\right] + \max\left[\frac{\left|\dot{x}_{3d} - f(\boldsymbol{x},t)\right| + \gamma}{\left|b(\boldsymbol{x},t)\right|}\right] = u_{max}$$

then the constant U specified above actually exists. Moreover, if

$$\left|\dot{e}_3(t)\right| \le \delta U, \tag{3.42}$$

then inequality (3.40) is satisfied. The analysis presented up to now is actually very similar to the one given at the beginning of section 2.2.1. The only difference is that now $\dot{x}_{2d}$ and $\dot{e}_{2d}$ are replaced with $\dot{x}_{3d}$ and $\dot{e}_{3d}$. Now we find the greatest value of $\left|\dot{e}_3(t)\right|$. We consider two different intervals: the first one from zero to t_f and the second one from t_f to infinity. For $t \le t_f$ the tracking error is given by equation (3.20). Differentiating this equation three times (or differentiating once equation (3.22)), we get

$$\dot{e}_3(t) = Be^{-\sqrt{c_1}t}\left(t\sqrt{c_1}-1\right).$$ (3.43)

Notice that at the initial time $\dot{e}_3(0) = -B$. Then we calculate the extreme value of this signal. The value is achieved at the time

$$t_{mu\,1} = \frac{2}{\sqrt{c_1}}$$ (3.44)

and it is equal to

$$\dot{e}_3(t_{mu\,1}) = \frac{B}{e^2}.$$ (3.45)

Note that $|Be^{-2}| < |B|$. Furthermore at the time $t = t_f$, $\dot{e}_3(t_f) = Be^{-k}(k-1)$. Notice that for $k > 0$ we have $|Be^{-k}(k-1)| \leq |B|$. This can be proved considering the right hand side of the above relation as a function of $k > 0$. Let $\varphi(k) = Be^{-k}(k-1)$. This function for $k = 0$ equals $\varphi(0) = -B$ and for $k \to \infty$, $\varphi(k) \to 0$. The extreme of this function is achieved when $k = 2$ and it is equal $\varphi(2) = Be^{-2}$. Then $|Be^{-2}| < |B|$. We conclude that for $t \leq t_f$ the greatest value of $|\dot{e}_3(t)|$ is equal to $|B|$.

On the other hand, for $t \geq t_f$ the trajectory following error is expressed by equation (3.28). Differentiating this equation three times (or differentiating once equation (3.30)) we get

$$\dot{e}_3(t) = -\sqrt{c_1}e^{-\sqrt{c_1}t}\left[-e_0e^k c_1 + B\left(e^k-1\right)t\right] + B\left(e^k-1\right)e^{-\sqrt{c_1}t}.$$ (3.46)

Consequently, at the time t_f

$$\dot{e}_3(t_f) = c_1\sqrt{c_1}\,e_0e^{-k} + B\left(e^k-1\right)e^{-k} = B\left[e^{-k}(k-1)+1\right].$$ (3.47)

The absolute value of the right hand side of the above equation may be greater than $|B|$ and could possibly present the greatest value of $|\dot{e}_3(t)|$. Furthermore, for $t \in \langle t_f, \infty\rangle$ the extreme value of (3.46) is reached at the time

$$t_{mu\,2} = \frac{2}{\sqrt{c_1}} + \frac{t_f e^k}{e^k-1}.$$ (3.48)

This extreme value is

$$\dot{e}_3(t_{mu\,2}) = -B\left(e^k-1\right)\exp\left[-\left(2+\frac{ke^k}{e^k-1}\right)\right].$$ (3.49)

3.2 Switching Plane Design Minimising IAE

Let us consider the following expression $|(e^k-1)\exp\{-[2+ke^k/(e^k-1)]\}| = |\exp[-2-k/(e^k-1)] - \exp[-2-ke^k/(e^k-1)]| < 1$. Therefore, $|B(e^k-1)\exp\{-[2+ke^k/(e^k-1)]\}| < |B|$. Finally, let us notice that the third derivative of the tracking error described by relation (3.46) converges to zero for $t \to \infty$. Consequently, we conclude that the greatest value of $|\dot{e}_3(t)|$ equals $|B|$ for $k < 1$ and is represented by the absolute value of expression (3.47) for $k \geq 1$. Therefore, we get two constraints: when $k < 1$ the constraint is expressed by inequality (2.55) and when $k \geq 1$ it is given by the following relation

$$|B| \leq \frac{\delta U}{e^{-k}(k-1)+1}. \quad (3.50)$$

These constraints are illustrated in figure 3.1. Now we will minimise criterion (3.37) subject to constraints (2.55) and (3.50). Since for any given value of k, criterion (3.37) is minimised for the greatest value of $|B|$ satisfying constraints (2.55) and (3.50), then the minimum of the criterion may be determined as a minimum of a single variable function $J_{IAE}^u(k)$. Consequently, the following two cases $k < 1$ and $k \geq 1$, should be considered.

Fig. 3.1 Optimisation constraints

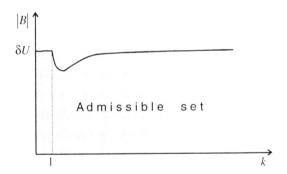

Taking into account constraints (2.55) and (3.50) we get the following control quality criterion as a function of a single variable

$$J_{IAE}^u(k) = \begin{cases} \dfrac{|e_0|^{4/3}}{(\delta U)^{1/3}} \left(2k^{-1/3} + \dfrac{1}{2}k^{2/3}\right) & \text{for } k < 1, \\[2ex] \dfrac{|e_0|^{4/3}\left[e^{-k}(k-1)+1\right]^{1/3}}{(\delta U)^{1/3}} \left(2k^{-1/3} + \dfrac{1}{2}k^{2/3}\right) & \text{for } k \geq 1. \end{cases} \quad (3.51)$$

This criterion is shown in figure 3.2.

Fig. 3.2 Criterion $J^u_{IAE}(k)$

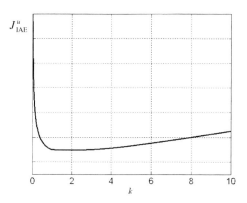

Then we calculate the derivative of criterion (3.51)

$$\frac{dJ^u_{IAE}(k)}{dk} = \begin{cases} \dfrac{|e_0|^{4/3}}{3(\delta U)^{1/3}} k^{-4/3}(-2+k) & \text{for } k<1, \\ \dfrac{|e_0|^{4/3}}{3(\delta U)^{1/3}} \left[e^{-k}(k-1)+1\right]^{-2/3} k^{-4/3} e^{-k} \\ \quad \cdot (-2+k)\left(e^k - 1 - k - \dfrac{1}{2}k^2\right) & \text{for } k \geq 1. \end{cases} \quad (3.52)$$

Notice that when $k < 2$ the above derivative is always negative. Furthermore, for $k > 2$ derivative (3.52) is always positive. Hence, we obtain the following optimal solution $k_{u\,opt} = 2$, where criterion (3.37) achieves its minimum value

$$J^u_{IAE}(k_{u\,opt}) = J^u_{IAE}(2) = \frac{3|e_0|^{4/3}}{(\delta U)^{1/3}} \left(\frac{e^{-2}+1}{2}\right)^{1/3}. \quad (3.53)$$

For $k = k_{u\,opt} = 2$, from the following formula

$$B = \text{sgn}(e_0) \frac{\delta U}{e^{-k}(k-1)+1}, \quad (3.54)$$

we get the optimal value of parameter B

$$B_{u\,opt} = \text{sgn}(e_0) \frac{\delta U}{e^{-k_{u\,opt}}(k_{u\,opt}-1)+1} = \text{sgn}(e_0) \frac{\delta U}{e^{-2}+1} \quad (3.55)$$

3.2 Switching Plane Design Minimising IAE

which concludes the minimisation of criterion (3.37) subject to constraint (2.47). Finally, the optimal switching plane parameters can be determined from relations (3.36), (3.15) and (3.19). These parameters are given below

$$c_{1\,u\,opt} = \left[\frac{2\,\delta U}{|e_0|\left(e^{-2}+1\right)}\right]^{\frac{2}{3}}, \tag{3.56}$$

$$c_{2\,u\,opt} = 2\left[\frac{2\,\delta U}{|e_0|\left(e^{-2}+1\right)}\right]^{\frac{1}{3}}, \tag{3.57}$$

$$A_{u\,opt} = -e_0\left[\frac{2\,\delta U}{|e_0|\left(e^{-2}+1\right)}\right]^{\frac{2}{3}}. \tag{3.58}$$

The switching plane determined in this section moves until

$$t_{f\,u\,opt} = 2^{2/3}\left[\frac{|e_0|\left(e^{-2}+1\right)}{\delta U}\right]^{1/3}. \tag{3.59}$$

Further in this section, similarly as in section 2.2.1, the method of the switching plane design, which ensures good dynamic performance and satisfies elastic input constraint, is presented. In this method we minimise another criterion which is a sum of the IAE and some penalty function. This criterion, expressed by equation (2.63), determines the convergence rate of the system and at the same time penalizes excessive values of control signal $|u(t)|$. Taking into account relation (3.37), criterion (2.63) can be presented as follows

$$Q_{IAE}^{u} = \frac{|e_0|^{4/3}}{|B|^{1/3}}\left(2k^{-1/3}+\frac{1}{2}k^{2/3}\right)+q\left\{\frac{\max\left[|u(t)|\right]}{M}\right\}^{n}. \tag{3.60}$$

Replacing M with u_{max} we obtain

$$Q_{IAE}^{u} = \frac{|e_0|^{4/3}}{|B|^{1/3}}\left(2k^{-1/3}+\frac{1}{2}k^{2/3}\right)+q\left\{\frac{\max\left[|u(t)|\right]}{u_{max}}\right\}^{n}. \tag{3.61}$$

Since inequality (3.42) implies (2.47), then in order to design the switching plane we replace criterion (3.61) with

$$Q_{IAE}^u(k,|B|) = \frac{|e_0|^{4/3}}{|B|^{1/3}}\left(2k^{-1/3} + \frac{1}{2}k^{2/3}\right) + q\left\{\frac{\max\left[|\dot{e}_3(t)|\right]}{\delta U}\right\}^n. \tag{3.62}$$

Now we will minimise criterion (3.62). Since $\max\left[|\dot{e}_3(t)|\right]$ is given by different formulae when $k < 1$ and when $k \geq 1$, we will consider two cases in the minimisation procedure. We begin with the case when $k < 1$. Then from equation (3.62) and our previous considerations we obtain

$$Q_{IAE}^u(k,|B|) = \frac{|e_0|^{4/3}}{|B|^{1/3}}\left(2k^{-1/3} + \frac{1}{2}k^{2/3}\right) + q\left(\frac{|B|}{\delta U}\right)^n. \tag{3.63}$$

This criterion will be minimised as a two variable (k and $|B|$) function. Clearly it can be minimised by finding its partial derivatives with respect to both variables and equating these derivatives to zero. The derivatives are

$$\frac{\partial Q_{IAE}^u(k,|B|)}{\partial k} = \frac{|e_0|^{4/3}}{|B|^{1/3}}\left(-\frac{2}{3}k^{-4/3} + \frac{1}{3}k^{-1/3}\right) = \frac{|e_0|^{4/3}}{3|B|^{1/3}k^{4/3}}(-2+k), \tag{3.64}$$

$$\frac{\partial Q_{IAE}^u(k,|B|)}{\partial |B|} = -\frac{|e_0|^{4/3}}{3|B|^{4/3}}\left(2k^{-1/3} + \frac{1}{2}k^{2/3}\right) + qn\frac{|B|^{n-1}}{(\delta U)^n}. \tag{3.65}$$

Notice that for $k < 1$ the right hand side of (3.64) is always negative. Consequently, criterion (3.62) decreases with increasing k. Hence we analyse the situation when $k \to 1$. Furthermore, solving $\partial Q_{IAE}^u(k,|B|)/\partial |B| = 0$, from the following equation

$$|B|^{n+\frac{1}{3}} = \frac{|e_0|^{4/3}(\delta U)^n}{3qn}\left(2k^{-1/3} + \frac{1}{2}k^{2/3}\right), \tag{3.66}$$

we obtain the formula describing parameter $|B|$ whose value for $k \to 1$ equals

$$|B_1| = \left[\frac{5|e_0|^{4/3}(\delta U)^n}{6qn}\right]^{\frac{3}{3n+1}}. \tag{3.67}$$

In this case criterion (3.62) reaches the following minimum value

$$Q_{IAE}^u(1,B_1) = \frac{5|e_0|^{4/3}(6qn)^{\frac{1}{3n+1}}}{2\left[5|e_0|^{4/3}(\delta U)^n\right]^{\frac{1}{3n+1}}} + \frac{q}{(\delta U)^n}\left[\frac{5|e_0|^{4/3}(\delta U)^n}{6qn}\right]^{\frac{3n}{3n+1}}. \tag{3.68}$$

3.2 Switching Plane Design Minimising IAE

In the second case, i.e. when $k \geq 1$ we minimise the criterion

$$Q_{\text{IAE}}^{u}\left(k,\left|B\right|\right) = \frac{\left|e_0\right|^{4/3}}{\left|B\right|^{1/3}}\left(2k^{-1/3} + \frac{1}{2}k^{2/3}\right) + q\left\{\frac{\left|B\right|\left[e^{-k}\left(k-1\right)+1\right]}{\delta U}\right\}^{n}. \tag{3.69}$$

In order to find the minimum of criterion (3.69), we calculate again the partial derivatives of $Q_{\text{IAE}}^{u}\left(k,\left|B\right|\right)$, getting in this way

$$\frac{\partial Q_{\text{IAE}}^{u}\left(k,\left|B\right|\right)}{\partial k} = \frac{\left|e_0\right|^{4/3}}{3\left|B\right|^{1/3}k^{4/3}}\left(-2+k\right) + $$
$$+ qn\left(\frac{\left|B\right|}{\delta U}\right)^{n}\left[e^{-k}\left(k-1\right)+1\right]^{n-1}e^{-k}\left(2-k\right), \tag{3.70}$$

$$\frac{\partial Q_{\text{IAE}}^{u}\left(k,\left|B\right|\right)}{\partial \left|B\right|} = -\frac{\left|e_0\right|^{4/3}}{3\left|B\right|^{4/3}}\left(2k^{-1/3} + \frac{1}{2}k^{2/3}\right) + $$
$$+ qn\frac{\left|B\right|^{n-1}}{\left(\delta U\right)^{n}}\left[e^{-k}\left(k-1\right)+1\right]^{n}. \tag{3.71}$$

Solving equation $\partial Q_{\text{IAE}}^{u}\left(k,\left|B\right|\right)/\partial\left|B\right| = 0$, i.e. equation

$$-\frac{\left|e_0\right|^{4/3}}{3\left|B\right|^{4/3}}\left(2k^{-1/3} + \frac{1}{2}k^{2/3}\right) + qn\frac{\left|B\right|^{n-1}}{\left(\delta U\right)^{n}}\left[e^{-k}\left(k-1\right)+1\right]^{n} = 0, \tag{3.72}$$

we obtain

$$\left|B\right|^{n+\frac{1}{3}} = \frac{\left|e_0\right|^{4/3}\left(\delta U\right)^{n}}{3qn\left[e^{-k}\left(k-1\right)+1\right]^{n}}\left(2k^{-1/3} + \frac{1}{2}k^{2/3}\right). \tag{3.73}$$

Therefore,

$$\left|B\right| = \left\{\frac{\left|e_0\right|^{4/3}\left(\delta U\right)^{n}}{3qn\left[e^{-k}\left(k-1\right)+1\right]^{n}}\left(2k^{-1/3} + \frac{1}{2}k^{2/3}\right)\right\}^{\frac{3}{3n+1}}. \tag{3.74}$$

On the other hand from equation $\partial Q_{\text{IAE}}^{u}\left(k,\left|B\right|\right)/\partial k = 0$, which is equivalent to

$$\frac{|e_0|^{4/3}}{3|B|^{1/3} k^{4/3}}(-2+k)+qn\left(\frac{|B|}{\delta U}\right)^n \left[e^{-k}(k-1)+1\right]^{n-1} e^{-k}(2-k)=0 \qquad (3.75)$$

and from formula (3.74), we get the following equation

$$(k-2)\left[e^{-k}(k-1)+1-2ke^{-k}-\frac{1}{2}e^{-k}k^2\right]=0 \qquad (3.76)$$

whose solution is $k = 2$. Therefore, we may expect that criterion (3.69) achieves its minimum for $k = 2$ and parameter $|B|$

$$|B_2| = \left[\frac{|e_0|^{4/3}(\delta U)^n}{2^{1/3} qn (e^{-2}+1)^n}\right]^{\frac{3}{3n+1}}. \qquad (3.77)$$

In order to prove this fact, we calculate the second order derivatives of criterion $Q_{IAE}^u (k,|B|)$

$$\frac{\partial^2 Q_{IAE}^u (k,|B|)}{\partial k \partial k} = \frac{|e_0|^{4/3}}{|B|^{1/3}}\left(\frac{8}{9}k^{-7/3}-\frac{1}{9}k^{-4/3}\right)+qn\left(\frac{|B|}{\delta U}\right)^n \cdot$$
$$\cdot\left\{(n-1)\left[e^{-k}(k-1)+1\right]^{n-2}\left[e^{-k}(2-k)\right]^2+\left[e^{-k}(k-1)+1\right]^{n-1}e^{-k}(k-3)\right\}, \qquad (3.78)$$

$$\frac{\partial^2 Q_{IAE}^u (k,|B|)}{\partial k \partial |B|} = \frac{\partial^2 Q_{IAE}^u (k,|B|)}{\partial |B| \partial k} =$$
$$= -\frac{|e_0|^{4/3}}{9|B|^{4/3} k^{4/3}}(-2+k)+qn^2\frac{|B|^{n-1}}{(\delta U)^n}\left[e^{-k}(k-1)+1\right]^{n-1}e^{-k}(2-k), \qquad (3.79)$$

$$\frac{\partial^2 Q_{IAE}^u (k,|B|)}{\partial |B| \partial |B|} = \frac{4|e_0|^{4/3}}{9|B|^{7/3}}\left(2k^{-1/3}+\frac{1}{2}k^{2/3}\right)+$$
$$+qn(n-1)\frac{|B|^{n-2}}{(\delta U)^n}\left[e^{-k}(k-1)+1\right]^n \qquad (3.80)$$

and then, for $k = 2$ and $|B| = |B_2|$ we construct the matrix of second order derivatives

3.2 Switching Plane Design Minimising IAE

$$H = \begin{bmatrix} \dfrac{\partial^2 Q_{\mathrm{IAE}}^u \left(k,|B|\right)}{\partial k \partial k} \Big|_{\substack{k=2 \\ |B|=|B_2|}} & \dfrac{\partial^2 Q_{\mathrm{IAE}}^u \left(k,|B|\right)}{\partial k \partial |B|} \Big|_{\substack{k=2 \\ |B|=|B_2|}} \\[4mm] \dfrac{\partial^2 Q_{\mathrm{IAE}}^u \left(k,|B|\right)}{\partial |B| \partial k} \Big|_{\substack{k=2 \\ |B|=|B_2|}} & \dfrac{\partial^2 Q_{\mathrm{IAE}}^u \left(k,|B|\right)}{\partial |B| \partial |B|} \Big|_{\substack{k=2 \\ |B|=|B_2|}} \end{bmatrix}. \tag{3.81}$$

Substituting relations (3.78), (3.79) and (3.80) into (3.81), for $k = 2$ and $|B| = |B_2|$ we find

$$H = \begin{bmatrix} h_{11} & 0 \\ 0 & h_{22} \end{bmatrix} \tag{3.82}$$

where

$$h_{11} = \left(\frac{|e_0|^{4/3}}{2^{1/3}} \right)^{\frac{3n}{3n+1}} \left[\frac{qn\left(e^{-2}+1\right)^n}{\left(\delta U\right)^n} \right]^{\frac{1}{3n+1}} \left(\frac{1}{6} - \frac{e^{-2}}{e^{-2}+1} \right), \tag{3.83}$$

$$h_{22} = \left(\frac{|e_0|^{4/3}}{2^{1/3}} \right)^{\frac{3n-6}{3n+1}} \left[qn \frac{\left(e^{-2}+1\right)^n}{\left(\delta U\right)^n} \right]^{\frac{7}{3n+1}} \left(n + \frac{1}{3} \right). \tag{3.84}$$

Since the following two inequalities always hold

$$\left(\frac{|e_0|^{4/3}}{2^{1/3}} \right)^{\frac{3n}{3n+1}} \left[\frac{qn\left(e^{-2}+1\right)^n}{\left(\delta U\right)^n} \right]^{\frac{1}{3n+1}} \left(\frac{1}{6} - \frac{e^{-2}}{e^{-2}+1} \right) > 0, \tag{3.85}$$

$$\left(\frac{|e_0|^{4/3}}{2^{1/3}} \right)^{\frac{3n-6}{3n+1}} \left[qn \frac{\left(e^{-2}+1\right)^n}{\left(\delta U\right)^n} \right]^{\frac{7}{3n+1}} \left(n + \frac{1}{3} \right) > 0, \tag{3.86}$$

we conclude that $\det(H) > 0$ and matrix H is positive definite. Therefore, for $k = 2$ and

$$B_2 = \mathrm{sgn}\left(e_0\right) \left[\frac{|e_0|^{4/3}\left(\delta U\right)^n}{2^{1/3} qn\left(e^{-2}+1\right)^n} \right]^{\frac{3}{3n+1}} \tag{3.87}$$

the considered criterion reaches its minimum, which is given by the following formula

$$Q^u_{\text{IAE}}(2, B_2) = \frac{3|e_0|^{4/3}\left[2^{1/3}\, qn\left(e^{-2}+1\right)^n\right]^{\frac{1}{3n+1}}}{2^{1/3}\left[|e_0|^{4/3}\left(\delta U\right)^n\right]^{\frac{1}{3n+1}}} +$$

$$+\frac{q\left(e^{-2}+1\right)^n}{\left(\delta U\right)^n}\left[\frac{|e_0|^{4/3}\left(\delta U\right)^n}{2^{1/3}\, qn\left(e^{-2}+1\right)^n}\right]^{\frac{3n}{3n+1}}. \tag{3.88}$$

Furthermore, the difference

$$Q^u_{\text{IAE}}(1, B_1) - Q^u_{\text{IAE}}(2, B_2) =$$

$$= \frac{|e_0|^{4/3}\left(qn\right)^{1/(3n+1)}}{\left[|e_0|^{4/3}\left(\delta U\right)^n\right]^{1/(3n+1)}}\left\{\frac{5}{2}\left(\frac{6}{5}\right)^{\frac{1}{3n+1}} - \frac{3\left[2^{1/3}\left(e^{-2}+1\right)^n\right]^{\frac{1}{3n+1}}}{2^{1/3}}\right\} + \tag{3.89}$$

$$+\frac{q}{\left(\delta U\right)^n}\left[\frac{|e_0|^{4/3}\left(\delta U\right)^n}{qn}\right]^{3n/(3n+1)}\left\{\left(\frac{5}{6}\right)^{\frac{3n}{3n+1}} - \frac{\left(e^{-2}+1\right)^n}{\left[2^{1/3}\left(e^{-2}+1\right)^n\right]^{\frac{3n}{3n+1}}}\right\},$$

and after some calculations we get

$$Q^u_{\text{IAE}}(1, B_1) - Q^u_{\text{IAE}}(2, B_2) =$$

$$= \left\{3\frac{|e_0|^{4/3}\left(qn\right)^{\frac{1}{3n+1}}}{\left[|e_0|^{4/3}\left(\delta U\right)^n\right]^{\frac{1}{3n+1}}} + \frac{q}{\left(\delta U\right)^n}\left[\frac{|e_0|^{4/3}\left(\delta U\right)^n}{qn}\right]^{\frac{3n}{3n+1}}\right\}. \tag{3.90}$$

$$\cdot\left[\left(\frac{5^3}{6^3}\right)^{\frac{n}{3n+1}} - \left(\frac{e^{-2}+1}{2}\right)^{\frac{n}{3n+1}}\right].$$

Since q, n, U and δ are strictly positive, then $Q^u_{\text{IAE}}(1, B_1) - Q^u_{\text{IAE}}(2, B_2) > 0$. Consequently, $Q^u_{\text{IAE}}(2, B_2)$ is the minimum value of the considered criterion in both cases ($k < 1$ and $k \geq 1$). This leads to the conclusion that parameters $k = 2$ and B_2 described by (3.87) are the optimal (in the sense of criterion (3.62)) switching

3.2 Switching Plane Design Minimising IAE

plane parameters. Criterion $Q_{IAE}^u(k,|B|)$ is shown in figure 3.3. The other switching plane parameters are given as follows

$$c_{1u\,opt} = \left\{ \frac{2}{|e_0|} \left[\frac{|e_0|^{4/3}(\delta U)^n}{2^{1/3}qn(e^{-2}+1)^n} \right]^{\frac{3}{3n+1}} \right\}^{\frac{2}{3}}, \qquad (3.91)$$

$$c_{2u\,opt} = 2 \left\{ \frac{2}{|e_0|} \left[\frac{|e_0|^{4/3}(\delta U)^n}{2^{1/3}qn(e^{-2}+1)^n} \right]^{\frac{3}{3n+1}} \right\}^{\frac{1}{3}}, \qquad (3.92)$$

$$A_{u\,opt} = -e_0 \left\{ \frac{2}{|e_0|} \left[\frac{|e_0|^{4/3}(\delta U)^n}{2^{1/3}qn(e^{-2}+1)^n} \right]^{\frac{3}{3n+1}} \right\}^{\frac{2}{3}}. \qquad (3.93)$$

The plane stops moving and becomes fixed at the time

$$t_{f\,u\,opt} = 2^{2/3}|e_0|^{1/3} \left[\frac{2^{1/3}qn(e^{-2}+1)^n}{|e_0|^{4/3}(\delta U)^n} \right]^{\frac{1}{3n+1}}. \qquad (3.94)$$

Similarly as in section 2.2.1 putting $n \to \infty$ into above equations and calculating appropriate limits of (3.87), (3.91), (3.92), (3.93) and (3.94) we get the switching plane parameters expressed by (3.55), (3.56), (3.57), (3.58) and (3.59), respectively.

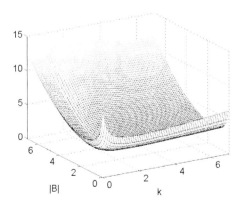

Fig. 3.3 The modified criterion $Q_{IAE}^u(k,|B|)$

Fig. 3.4 Tracking error evolution and its derivatives (conventional input constraint)

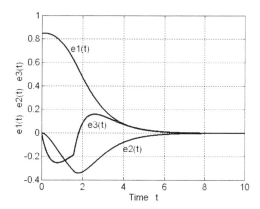

Fig. 3.5 Control signal (conventional input constraint)

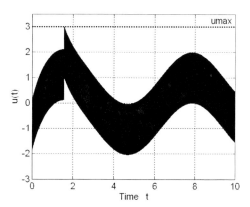

Fig. 3.6 Tracking error evolution and its derivatives ($n = 5$, $q = 5$)

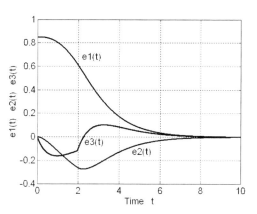

3.2 Switching Plane Design Minimising IAE

Fig. 3.7 Control signal ($n = 5$, $q = 5$)

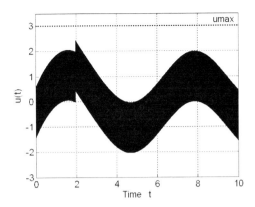

Fig. 3.8 Tracking error evolution and its derivatives ($n = 5$, $q = 0.02$)

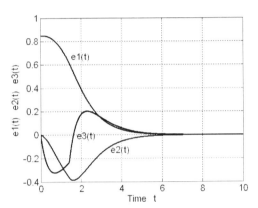

Fig. 3.9 Control signal ($n = 5$, $q = 0.02$)

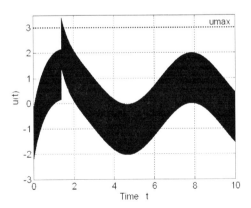

Simulation example

In order to verify the system performance when the input signal constraint is required to be satisfied we consider the following triple integrator

$$\dot{x}_1 = x_2$$
$$\dot{x}_2 = x_3 \qquad\qquad (3.95)$$
$$\dot{x}_3 = \Delta f (x,t) + d(t) + u$$

with model uncertainty

$$\Delta f(x,t) = 0.39 \sin\left(x_1 x_2 + x_3 \sqrt{t}\right) \qquad\qquad (3.96)$$

and external disturbance given by equality (2.86). Consequently $\gamma = 1$. The initial conditions are specified as follows

$$x_{10} = 1.85, \;\; x_{20} = 0, \;\; x_{30} = -1. \qquad\qquad (3.97)$$

System (3.95) is supposed to track the demand trajectory

$$x_{1d}(t) = \cos t. \qquad\qquad (3.98)$$

We require the following condition $|u(t)| \le u_{max} = 3$ to be satisfied, which leads to the conclusion that $U = 1$. Using the method presented in this section we obtain $c_1 \approx 1.62$, $c_2 \approx 2.55$, $B \approx 0.88$, $A \approx -1.38$. The switching plane stops moving at the time instant $t_f \approx 1.57$. The tracking error evolution and its derivatives are presented in figure 3.4. From this figure it can be seen that the system is insensitive from the very beginning of the control process. Figure 3.5 shows that the input signal constraint is always satisfied. Moreover, figures 3.6 – 3.9 illustrate the system performance when elastic constraint is considered. We take into account two cases when $n = 5$, $q = 5$ and $n = 5$, $q = 0.02$. For $n = 5$ and $q = 5$ the optimal switching plane parameters are given as follows $c_1 \approx 1.033$, $c_2 \approx 2.033$, $B \approx 0.446$, $A \approx -0.878$, $t_f \approx 1.969$. For $n = 5$, $q = 0.02$ we obtain $c_1 \approx 2.0605$, $c_2 \approx 2.871$, $B \approx 1.257$, $A \approx -1.751$, $t_f \approx 1.393$. From figures 3.6 – 3.9 it can be seen not only that the system is insensitive with respect to external disturbance and model uncertainty but also how different selection of parameter q affects the maximum value of the input signal.

3.2.2 Switching Plane Design Subject to Acceleration Constraint

In this section we introduce the acceleration constraint, i.e. we require the system acceleration not to exceed the maximum admissible value a_{max}

$$\left|e_3(t)\right| \le a_{max} . \qquad\qquad (3.99)$$

3.2 Switching Plane Design Minimising IAE

Now we find the maximum absolute value of the system acceleration. Equation (3.22) represents the system acceleration for $t \leq t_f$, that is before the switching plane stops moving. The extreme value of this function is achieved at

$$t_{ma\,1} = \frac{1}{\sqrt{c_1}} \tag{3.100}$$

and it is equal to

$$e_3\,(t_{ma\,1}) = -\frac{B}{e\sqrt{c_1}}. \tag{3.101}$$

Further in this section, we consider two cases: one when $t_{ma\,1} < t_f$ and the other when $t_{ma\,1} \geq t_f$.

Case 1 $(t_{ma\,1} < t_f)$**:** If $t_{ma\,1} < t_f$, then the absolute value of the right-hand side of equation (3.101) is the greatest acceleration/deceleration of the system both when the plane moves and after it stops. In order to show this property, it is necessary to compare this quantity with the extreme values of the function on the right-hand side of equation (3.30). That function reaches its extreme value, equal

$$e_3(t_{ma\,2}) = \exp[-ke^k/(e^k-1)](e^k-1)B/\,e\,\sqrt{c_1} \tag{3.102}$$

at the time instant

$$t_{ma\,2} = 1/\sqrt{c_1} + e_0 e^k c_1/B(e^k-1). \tag{3.103}$$

Since $k > 0$, then $\exp[-ke^k/(e^k-1)](e^k-1) < 1$. Consequently, inequality $|e_3(t_{ma\,1})| > |e_3(t_{ma\,2})|$ is always satisfied. Therefore, the absolute value of $e_3(t_{ma\,1})$ is actually the greatest acceleration of the considered system for any time $t \geq 0$. As a result, we get the following constraint for the considered minimisation of criterion (3.37)

$$|B| \leq a_{max}\,e\,\sqrt{c_1} \tag{3.104}$$

where a_{max} is the maximum admissible value of the system acceleration. Taking into account equation (3.36) we get the following form of constraint (3.104)

$$|B| \leq \left(\frac{a_{max}\,ek^{1/3}}{|e_0|^{1/3}}\right)^{3/2}. \tag{3.105}$$

Notice that in the analysed case $t_{ma1} < t_f$ and consequently,

$$\frac{c_1 e_0}{B} > \frac{1}{\sqrt{c_1}}. \tag{3.106}$$

This is equivalent to the situation when $k > 1$. Thus we conclude that for any $k > 1$ the acceleration constraint is expressed by inequality (3.105).

Case 2 ($t_f \leq t_{ma1}$): In this case we have $|B| \geq |e_0|\sqrt{c_1}c_1$ which is equivalent to $k \leq 1$. Now the greatest system acceleration/deceleration is given by the absolute value of (3.27). Therefore, the explicit constraint formulation can be found from

$$a_{max} \geq e^{-k}c_1|e_0|. \tag{3.107}$$

From this relation and equation (3.36) we get

$$|B| \leq \left(\frac{a_{max}}{e^{-k}|e_0|^{1/3} k^{2/3}} \right)^{3/2}. \tag{3.108}$$

This inequality represents the acceleration constraint for any $k \leq 1$. Taking into account constraints (3.105) and (3.108), and substituting them into (3.37) we get the following form of the considered criterion as a function of a single variable

$$J_{IAE}^a(k) = \begin{cases} \dfrac{|e_0|^{3/2}}{\sqrt{a_{max}}} \left(2e^{-k/2} + \dfrac{k}{2}e^{-k/2} \right) & \text{for } k \leq 1, \\ \dfrac{|e_0|^{3/2}}{\sqrt{a_{max}}\,e} \left(2k^{-1/2} + \dfrac{1}{2}k^{1/2} \right) & \text{for } k > 1. \end{cases} \tag{3.109}$$

Criterion (3.109) is illustrated in figure 3.10.

Fig. 3.10 Criterion $J_{IAE}^a(k)$

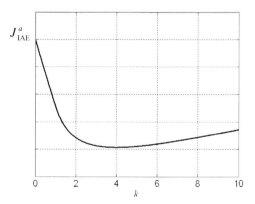

3.2 Switching Plane Design Minimising IAE

Calculating the derivative of this criterion we obtain

$$\frac{dJ^a_{IAE}(k)}{dk} = \begin{cases} \dfrac{-\left|e_0\right|^{3/2} e^{-k/2}}{2\sqrt{a_{max}}}\left(1+\dfrac{k}{2}\right) & \text{for } k \le 1, \\[4mm] \dfrac{\left|e_0\right|^{3/2}}{\sqrt{a_{max}}\,ek}\left(-k^{-1}+\dfrac{1}{4}\right) & \text{for } k > 1. \end{cases} \tag{3.110}$$

Checking the sign of this derivative we conclude that $J^a_{IAE}(k)$ decreases for $k \in (0, 4)$ and it increases for any $k > 4$. Thus, for $k_{a\,opt} = 4$ this function reaches its minimum

$$J^a_{IAE}\left(k_{a\,opt}\right) = J^a_{IAE}(4) = \frac{2\left|e_0\right|^{3/2}}{\sqrt{a_{max}}\,e}, \tag{3.111}$$

and then from the following formula

$$B = \text{sgn}\left(e_0\right)\left(\frac{a_{max}\,ek^{1/3}}{\left|e_0\right|^{1/3}}\right)^{3/2}, \tag{3.112}$$

we get the optimal value of parameter B

$$B_{a\,opt} = \text{sgn}\left(e_0\right)\left(\frac{a_{max}\,ek^{1/3}_{a\,opt}}{\left|e_0\right|^{1/3}}\right)^{3/2} = 2\,\text{sgn}\left(e_0\right)\left(\frac{a_{max}\,e}{\left|e_0\right|^{1/3}}\right)^{3/2}. \tag{3.113}$$

The pair $(k_{a\,opt}, B_{a\,opt})$ is the optimal solution of the criterion $J_{IAE}(k)$ minimisation task subject to the acceleration constraint. Having calculated parameters k and B, we can obtain the other switching plane parameters

$$c_{1\,a\,opt} = \frac{4\,a_{max}\,e}{\left|e_0\right|} \tag{3.114}$$

$$c_{2\,a\,opt} = 4\sqrt{\frac{a_{max}\,e}{\left|e_0\right|}} \tag{3.115}$$

$$A_{a\,opt} = -4\,\text{sgn}\left(e_0\right)a_{max}\,e. \tag{3.116}$$

Then the plane stops moving at the time instant

$$t_{f\ a\ opt} = 2\sqrt{\frac{|e_0|}{a_{\max}\ e}}. \tag{3.117}$$

Now we will modify the analysed criterion and we will present the method of the switching plane design, which ensures good dynamic performance and satisfies elastic acceleration constraint. In this method the following control quality criterion is minimised

$$Q^a_{\text{IAE}} = \int_{t_0}^{\infty} |e_1(t)| dt + q\left\{\frac{\max\left[|e_3(t)|\right]}{M}\right\}^n \tag{3.118}$$

where this time $M > 0$ is the threshold value of the acceleration, $q > 0$ is, as previously, a weighting factor and $n \geq 1$, as in the previous section, is a constant determining how elastic (or stretchable) the constraint is. Our earlier considerations make us expect that also in this case for $n \to \infty$ and M taken as $a_{\max}$, minimising criterion (3.118) we will get the control law which guarantees the minimum integral of the absolute error strictly without exceeding the maximum admissible value of the acceleration.

Assuming that $a_{\max}$ is the threshold value of the system acceleration and taking into account (3.37), we get

$$Q^a_{\text{IAE}} = \frac{|e_0|^{4/3}}{|B|^{1/3}}\left(2\,k^{-1/3} + \frac{1}{2}k^{2/3}\right) + q\left\{\frac{\max\left[|e_3(t)|\right]}{a_{\max}}\right\}^n. \tag{3.119}$$

Now we will precisely analyse the minimisation task of criterion (3.119). We will consider the following two situations: first when $k \leq 1$ and the second one when $k > 1$. We begin with the case when $k \leq 1$. In this situation, since the maximum value of $e_3(t)$ is described by (3.27), criterion (3.119) can be presented as

$$Q^a_{\text{IAE}}(k,|B|) = \frac{|e_0|^{4/3}}{|B|^{1/3}}\left(2\,k^{-1/3} + \frac{1}{2}k^{2/3}\right) + q\left(\frac{|B|^{2/3}\ k^{2/3}\ |e_0|^{1/3}\ e^{-k}}{a_{\max}}\right)^n. \tag{3.120}$$

Performing the minimisation procedure of two variable function $Q^a_{\text{IAE}}(k,|B|)$, we begin with calculating its first order derivatives

$$\frac{\partial Q^a_{\text{IAE}}(k,|B|)}{\partial k} = \frac{|e_0|^{4/3}}{3|B|^{1/3}\ k^{4/3}}(-2+k) +$$

$$+ qn\left(\frac{|e_0|^{1/3}\ |B|^{2/3}}{a_{\max}}\right)^n \left(k^{2/3}\ e^{-k}\right)^n \left(\frac{2}{3}k^{-1} - 1\right), \tag{3.121}$$

3.2 Switching Plane Design Minimising IAE

$$\frac{\partial Q_{IAE}^a\left(k,|B|\right)}{\partial|B|} = -\frac{|e_0|^{4/3}}{3|B|^{4/3}}\left(2k^{-1/3}+\frac{1}{2}k^{2/3}\right)+\frac{2}{3}qn|B|^{2\,n/3-1}\left(\frac{|e_0|^{1/3}\,k^{2/3}}{a_{max}\,e^k}\right)^n . \quad (3.122)$$

Next solving equation $\partial Q_{IAE}^a\left(k,|B|\right)\big/\partial|B|=0$, we obtain

$$|B| = \left[\frac{|e_0|^{(4-n)/3}\left(2+k/2\right)\left(e^k\,a_{max}\right)^n\left(k^{1/3}\right)^{-2\,n-1}}{2\,qn}\right]^{\frac{3}{2\,n+1}} . \quad (3.123)$$

Substituting $|B|$ given by (3.123) into equation $\partial Q_{IAE}^a\left(k,|B|\right)\big/\partial k=0$, we obtain that for $k\le 1$ this equation does not have any solution. Moreover, criterion (3.120) decreases with increasing k. Then substituting $k=1$ into (3.123) we get

$$|B_1| = \left[\frac{5|e_0|^{(4-n)/3}\left(ea_{max}\right)^n}{4\,qn}\right]^{\frac{3}{2\,n+1}} . \quad (3.124)$$

In this case criterion (3.119) reaches its minimum value equal

$$Q_{IAE}^a\left(1,B_1\right) = \left(5|e_0|^{4/3}\right)^{\frac{2\,n}{2\,n+1}}\left[\frac{qn|e_0|^{n/3}}{\left(ea_{max}\right)^n}\right]^{\frac{1}{2\,n+1}} 2^{\frac{1-2\,n}{2\,n+1}} +$$

$$+q\left(\frac{|e_0|^{1/3}}{ea_{max}}\right)^n\left[\frac{5|e_0|^{(4-n)/3}\left(ea_{max}\right)^n}{4\,qn}\right]^{\frac{2\,n}{2\,n+1}} . \quad (3.125)$$

In the second situation, i.e. when $k>1$, the considered criterion has the following form

$$Q_{IAE}^a\left(k,|B|\right) = \frac{|e_0|^{4/3}}{|B|^{1/3}}\left(2k^{-1/3}+\frac{k^{2/3}}{2}\right)+q\left(\frac{|e_0|^{1/3}|B|^{2/3}\,k^{-1/3}}{ea_{max}}\right)^n . \quad (3.126)$$

Calculating the first derivatives of $Q_{IAE}^a\left(k,|B|\right)$

$$\frac{\partial Q_{IAE}^a\left(k,|B|\right)}{\partial k} = \frac{|e_0|^{4/3}}{3|B|^{1/3}\,k^{4/3}}\left(-2+k\right)-\frac{1}{3}qn\left(\frac{|e_0|^{1/3}|B|^{2/3}}{ea_{max}}\right)^n k^{-n/3-1}, \quad (3.127)$$

$$\frac{\partial Q_{\text{IAE}}^a\left(k,|B|\right)}{\partial |B|} = -\frac{|e_0|^{4/3}}{3|B|^{4/3}}\left(2k^{-1/3}+\frac{1}{2}k^{2/3}\right)+\frac{2}{3}qn|B|^{2\,n/3-1}\left(\frac{|e_0|^{1/3}}{ea_{\max}\,k^{1/3}}\right)^{n} \tag{3.128}$$

and solving equation $\partial Q_{\text{IAE}}^a\left(k,|B|\right)/\partial |B| = 0$ we obtain

$$|B| = \left[\frac{|e_0|^{(4-n)/3}\left(2+k/2\right)\left(ea_{\max}\right)^{n}\left(k^{1/3}\right)^{n-1}}{2\,qn}\right]^{\frac{3}{2\,n+1}}. \tag{3.129}$$

Furthermore, from formula (3.129) and equation $\partial Q_{\text{IAE}}^a\left(k,|B|\right)/\partial k = 0$ we have $2+k/2 = 2(-2+k)$. Hence we get the solution: $k = 4$ and

$$|B_2| = 2\left[\frac{|e_0|^{(4-n)/3}\left(ea_{\max}\right)^{n}}{qn}\right]^{\frac{3}{2\,n+1}}. \tag{3.130}$$

This pair is expected to be the optimal solution of the minimisation task. In order to verify that this is truly the case, we show that the matrix of the second order derivatives of $Q_{\text{IAE}}^a\left(k,|B|\right)$

$$H = \begin{bmatrix} \left.\dfrac{\partial^2 Q_{\text{IAE}}^a\left(k,|B|\right)}{\partial k\partial k}\right|_{\substack{k=4 \\ |B|=|B_2|}} & \left.\dfrac{\partial^2 Q_{\text{IAE}}^a\left(k,|B|\right)}{\partial k\partial |B|}\right|_{\substack{k=4 \\ |B|=|B_2|}} \\[4mm] \left.\dfrac{\partial^2 Q_{\text{IAE}}^a\left(k,|B|\right)}{\partial |B|\partial k}\right|_{\substack{k=4 \\ |B|=|B_2|}} & \left.\dfrac{\partial^2 Q_{\text{IAE}}^a\left(k,|B|\right)}{\partial |B|\partial |B|}\right|_{\substack{k=4 \\ |B|=|B_2|}} \end{bmatrix} \tag{3.131}$$

for $k = 4$ and $|B| = |B_2|$ is positive definite. Substituting the following second order derivatives

$$\frac{\partial^2 Q_{\text{IAE}}^a\left(k,|B|\right)}{\partial k\partial k} = \frac{|e_0|^{4/3}}{9|B|^{1/3}}\left(8k^{-7/3}-k^{-4/3}\right)+ \\ +\frac{1}{3}qn\left(\frac{n}{3}+1\right)\left(\frac{|e_0|^{1/3}|B|^{2/3}}{a_{\max}\,e}\right)^{n}k^{-n/3-2}, \tag{3.132}$$

3.2 Switching Plane Design Minimising IAE

$$\frac{\partial^2 Q_{\text{IAE}}^a\left(k,|B|\right)}{\partial k \partial |B|} = \frac{\partial^2 Q_{\text{IAE}}^a\left(k,|B|\right)}{\partial |B| \partial k} =$$

$$= -\frac{|e_0|^{4/3}}{9|B|^{4/3} k^{4/3}}(-2+k) - \frac{2}{9}qn^2 |B|^{2\,n/3-1}\left(\frac{|e_0|^{1/3}}{a_{\max}\,e}\right)^n k^{-n/3-1}, \qquad (3.133)$$

$$\frac{\partial^2 Q_{\text{IAE}}^a\left(k,|B|\right)}{\partial |B| \partial |B|} = \frac{4|e_0|^{4/3}}{9|B|^{7/3}}\left(2k^{-1/3} + \frac{1}{2}k^{2/3}\right) +$$

$$+ \frac{2}{3}qn\left(\frac{2}{3}n-1\right)|B|^{2\,n/3-2}\left(\frac{|e_0|^{1/3}}{a_{\max}\,ek^{1/3}}\right)^n \qquad (3.134)$$

into matrix (3.131), for $k = 4$ and $|B| = |B_2|$, we obtain

$$H = \begin{bmatrix} \dfrac{|e_0|^{4/3}\,(qn)^{\frac{1}{2\,n+1}}\,(n+5)}{16 \cdot 9\left[|e_0|^{\frac{4-n}{3}}\,(ea_{\max})^n\right]^{\frac{1}{2\,n+1}}} & \dfrac{-|e_0|^{4/3}\,(qn)^{\frac{4}{2\,n+1}}\left(n+\dfrac{1}{2}\right)}{36\left[|e_0|^{\frac{4-n}{3}}\,(ea_{\max})^n\right]^{\frac{4}{2\,n+1}}} \\[3em] \dfrac{-|e_0|^{4/3}\,(qn)^{\frac{4}{2\,n+1}}\left(n+\dfrac{1}{2}\right)}{36\left[|e_0|^{\frac{4-n}{3}}\,(ea_{\max})^n\right]^{\frac{4}{2\,n+1}}} & \dfrac{|e_0|^{4/3}\,(qn)^{\frac{7}{2\,n+1}}\left(n+\dfrac{1}{2}\right)}{9\left[|e_0|^{\frac{4-n}{3}}\,(ea_{\max})^n\right]^{\frac{7}{2\,n+1}}} \end{bmatrix}. \qquad (3.135)$$

Notice that the following two inequalities are always satisfied

$$\frac{|e_0|^{4/3}\,(qn)^{\frac{1}{2\,n+1}}}{144\left[|e_0|^{(4-n)/3}\,(ea_{\max})^n\right]^{\frac{1}{2\,n+1}}}(n+5) > 0, \qquad (3.136)$$

$$\frac{|e_0|^{8/3}\,(qn)^{\frac{8}{2\,n+1}}}{16 \cdot 81\left[|e_0|^{(4-n)/3}\,(ea_{\max})^n\right]^{\frac{8}{2\,n+1}}}(n+5)\left(n+\frac{1}{2}\right) +$$

$$-\frac{|e_0|^{8/3}\,(qn)^{\frac{8}{2\,n+1}}}{81\left[|e_0|^{(4-n)/3}\,(ea_{\max})^n\right]^{\frac{8}{2\,n+1}}}\left(2^{-2}\,n+2^{-3}\right)^2 = \qquad (3.137)$$

$$= \frac{|e_0|^{8/3}\,(qn)^{\frac{8}{2\,n+1}}}{288\left[|e_0|^{(4-n)/3}\,(ea_{\max})^n\right]^{\frac{8}{2\,n+1}}}\left(n+\frac{1}{2}\right) > 0.$$

This leads to the conclusion that $\det(\boldsymbol{H}) > 0$ and consequently $\boldsymbol{H}$ is positive definite. Therefore, for $k_{a\,opt} = 4$ and

$$
B_{a\,opt} = B_2 = 2\,\mathrm{sgn}\left(e_0\right)\left[\frac{\left|e_0\right|^{(4-n)/3}\left(ea_{max}\right)^n}{qn}\right]^{\frac{3}{2n+1}},
\tag{3.138}
$$

criterion (3.126) achieves its minimum

$$
Q_{IAE}^a\left(4,B_2\right) = 2\left(\left|e_0\right|^{4/3}\right)^{\frac{2n}{2n+1}}\left[qn\left(\frac{\left|e_0\right|^{1/3}}{a_{max}\,e}\right)^n\right]^{\frac{1}{2n+1}} +
$$

$$
+q\left(\frac{\left|e_0\right|^{1/3}}{a_{max}\,e}\right)^n\left[\frac{\left|e_0\right|^{(4-n)/3}\left(a_{max}\,e\right)^n}{qn}\right]^{\frac{2n}{2n+1}}.
\tag{3.139}
$$

Since constants q, n and a_{max} are strictly positive, we get

$$
Q_{IAE}^a\left(1,B_1\right) - Q_{IAE}^a\left(4,B_2\right) =
$$

$$
= \left\{ 2\left(\left|e_0\right|^{4/3}\right)^{\frac{2n}{2n+1}}\left[qn\left(\frac{\left|e_0\right|^{1/3}}{a_{max}\,e}\right)^n\right]^{\frac{1}{2n+1}} + \right.
$$

$$
\left. +q\left(\frac{\left|e_0\right|^{1/3}}{a_{max}\,e}\right)^n\left[\frac{\left|e_0\right|^{(4-n)/3}\left(a_{max}\,e\right)^n}{qn}\right]^{\frac{2n}{2n+1}} \right\}\left[\left(\frac{5}{4}\right)^{\frac{2n}{2n+1}} - 1\right] > 0.
\tag{3.140}
$$

Consequently, $Q_{IAE}^a\left(4,B_2\right)$ is the minimum value of criterion (3.119) in both cases ($k \leq 1$ and $k > 1$). Hence, $k_{a\,opt} = 4$ and parameter $B_{a\,opt}$ given by (3.138), are the optimal solution of the considered minimisation task. In this case criterion $Q_{IAE}^a\left(k,\left|B\right|\right)$ is illustrated in figure 3.11.

The other switching plane parameters are given as follows

$$
c_{1\,a\,opt} = 4\left|e_0\right|^{-2/3}\left[\frac{\left|e_0\right|^{(4-n)/3}\left(ea_{max}\right)^n}{qn}\right]^{\frac{2}{2n+1}},
\tag{3.141}
$$

3.2 Switching Plane Design Minimising IAE

Fig. 3.11 The modified criterion $Q^a_{IAE}(k,|B|)$

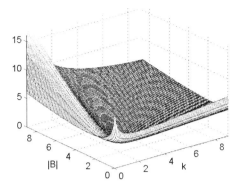

$$c_{2\,a\,opt} = 4|e_0|^{-1/3}\left[\frac{|e_0|^{(4-n)/3}(ea_{max})^n}{qn}\right]^{\frac{1}{2n+1}}, \qquad (3.142)$$

$$A_{a\,opt} = -4e_0|e_0|^{-2/3}\left[\frac{|e_0|^{(4-n)/3}(ea_{max})^n}{qn}\right]^{\frac{2}{2n+1}}, \qquad (3.143)$$

$$t_{f\,a\,opt} = 2|e_0|^{1/3}\left[\frac{|e_0|^{(4-n)/3}(ea_{max})^n}{qn}\right]^{\frac{-1}{2n+1}}. \qquad (3.144)$$

At the end of this subsection we analyse the situation when $n \to \infty$. Calculating appropriate limits of (3.138), (3.141), (3.142), (3.143) and (3.144) we get the switching plane parameters expressed by (3.113), (3.114), (3.115), (3.116) and (3.117), respectively. This shows that the conventional constraint considered at the beginning of the subsection is a special case of the elastic constraint analysed now.

Simulation example

In order to verify the system performance in the case considered in this section, i.e. when the acceleration is limited, we consider the following third order system

$$\begin{aligned}\dot{x}_1 &= x_2 \\ \dot{x}_2 &= x_3 \\ \dot{x}_3 &= -\exp(-x_3^2) + \frac{1}{2}\sin(x_1 x_2) + \Delta f(x,t) + d(t) + u\end{aligned} \qquad (3.145)$$

where relations (3.96) and (2.86) represent model uncertainty and external disturbance. Consequently we choose $\gamma = 1$. The initial conditions are

$$x_{10} = 0.85,\ x_{20} = -1,\ x_{30} = 0. \qquad (3.146)$$

Fig. 3.12 Tracking error evolution and its first derivative (conventional acceleration constraint)

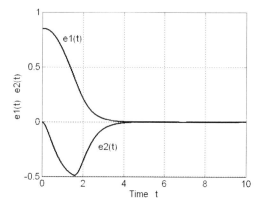

Fig. 3.13 Second derivative of the tracking error (conventional acceleration constraint)

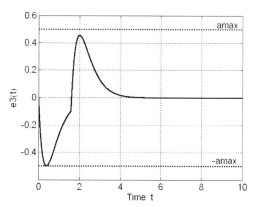

Fig. 3.14 Control signal (conventional acceleration constraint)

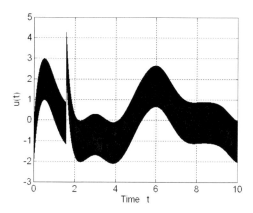

3.2 Switching Plane Design Minimising IAE

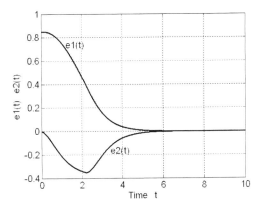

Fig. 3.15 Tracking error evolution and its first derivative ($n = 5$, $q = 5$)

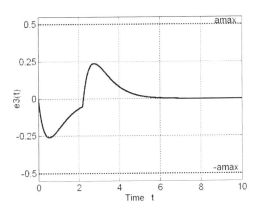

Fig. 3.16 Second derivative of the tracking error ($n = 5$, $q = 5$)

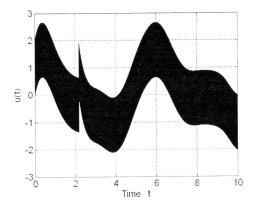

Fig. 3.17 Control signal ($n = 5$, $q = 5$)

Fig. 3.18 Tracking error evolution and its first derivative ($n = 5$, $q = 0.02$)

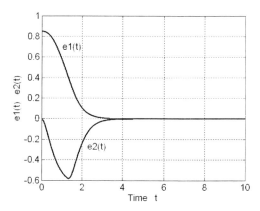

Fig. 3.19 Second derivative of the tracking error ($n = 5$, $q = 0.02$)

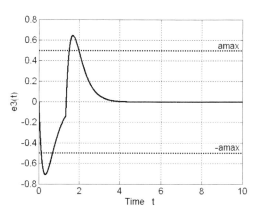

Fig. 3.20 Control signal ($n = 5$, $q = 0.02$)

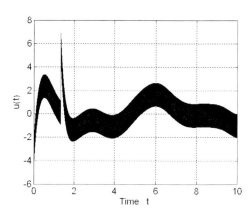

System (3.145) is supposed to track the demand trajectory

$$x_{1d}(t) = -\sin t. \qquad (3.147)$$

3.2 Switching Plane Design Minimising IAE

In this case we require that the maximum admissible value of the system accelera-tion is equal to $a_{max} = 0.5$. According to the presented method we get $c_1 \approx 6.395$, $c_2 \approx 5.058$, $B \approx 3.437$ and $A \approx -5.436$. The plane stops moving at time instant $t_f \approx 1.582$. The tracking error evolution and its first derivative are presented in fig-ure 3.12. Moreover, the second derivative of the tracking error is shown in figure 3.13. It can be seen from this figure that the imposed acceleration constraint is al-ways satisfied. Figure 3.14 illustrates the control signal in this case. Furthermore, the system performance when elastic constraint is considered can be seen from fig-ures 3.15 – 3.20. We take into account two cases: $n = 5$, $q = 5$ and $n = 5$, $q = 0.02$. For $n = 5$ and $q = 5$ the optimal switching plane parameters are $c_1 \approx 3.314$, $c_2 \approx 3.641$, $B \approx 1.282$, $A \approx -2.817$, $t_f \approx 2.197$. On the other hand, for $n = 5$, $q = 0.02$ we obtain $c_1 \approx 9.044$, $c_2 \approx 6.015$, $B \approx 5.78$, $A \approx -7.687$, $t_f \approx 1.33$. Figures 3.15 – 3.20 show that the system is insensitive with respect to external disturbance and model uncertainty. Moreover, these figures demonstrate how the different selection of pa-rameter q affects the maximum value of the system acceleration.

3.2.3 Switching Plane Design Subject to Velocity Constraint

In this section we will take into account system (3.1) subject to velocity constraint (2.89). Let us recall that for any time $t \le t_f$ the system velocity is described by equation (3.21) and for the time $t \ge t_f$ by equation (3.29). The extreme value of the velocity

$$e_2(t_{mv}) = \exp[-ke^k/(e^k - 1)](1 - e^k)B/c_1 \tag{3.148}$$

is achieved at the time instant $t_{mv} = e_0 c_1 e^k/B(e^k - 1)$. Notice that from relation (3.36) we have

$$e_2\left(t_{mv}\right) = \exp\left(\frac{-ke^k}{e^k - 1}\right)\left(1 - e^k\right)\frac{|B|^{1/3} e_0^{2/3}}{k^{2/3}} \operatorname{sgn}\left(e_0\right). \tag{3.149}$$

We require that

$$\left|\exp\left(\frac{-ke^k}{e^k - 1}\right)\left(1 - e^k\right)\frac{|B|^{1/3} e_0^{2/3}}{k^{2/3}} \operatorname{sgn}\left(e_0\right)\right| \le v_{max}. \tag{3.150}$$

Taking into account velocity constraint (3.150) we find the maximum admissible value of $|B|$

$$|B| \le \frac{v_{max}^3 k^2}{e_0^2 \left(e^k - 1\right)^3} \exp\left(\frac{3ke^k}{e^k - 1}\right). \tag{3.151}$$

Then substituting this value into criterion (3.37), similarly as in section 2.2.2, we obtain

$$J_{IAE}^v(k) = \frac{e_0^2}{v_{max}} \exp\left(\frac{-ke^k}{e^k - 1}\right)\left(\frac{2}{k} + \frac{1}{2}\right)(e^k - 1). \qquad (3.152)$$

This single variable function, shown in figure 3.21, for any k represents the minimum value of criterion $J_{IAE}(k, B)$ which can be achieved subject to velocity constraint (2.89).

Fig. 3.21 Criterion $J_{IAE}^v(k)$

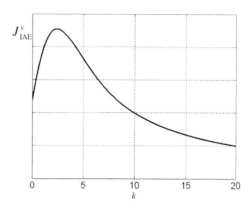

Now we introduce the following theorem.

Theorem 1
There exists such a value $k_{v\,max} \in (2, 2.5)$ that for any $k \in [0, k_{v\,max})$ function $J_{IAE}^v(k)$, expressed by (3.152), increases, and it decreases for every $k \in (k_{v\,max}, \infty)$.

Proof
Calculating the derivative of function $J_{IAE}^v(k)$, expressed by equation (3.152)

$$\frac{dJ_{IAE}^v(k)}{dk} = \frac{e_0^2}{v_{max}} \exp\left(\frac{-ke^k}{e^k - 1}\right) \frac{1}{2k^2(e^k - 1)} \left(4k^2 e^k + k^3 e^k + 8e^k - 4e^{2k} - 4\right) =$$

$$= \frac{e_0^2}{v_{max}} \exp\left(\frac{-ke^k}{e^k - 1}\right) \frac{e^k}{2k^2(e^k - 1)} \left(4k^2 + k^3 + 8 - 4e^k - 4e^{-k}\right), \qquad (3.153)$$

and using Euler's sequence, we obtain

3.2 Switching Plane Design Minimising IAE

$$\frac{dJ^v_{IAE}(k)}{dk} = \frac{e_0^2}{v_{max}} \exp\left(\frac{-ke^k}{e^k-1}\right) \frac{ke^k}{2(e^k-1)}\left(1 - 8\frac{k}{4!} - 8\frac{k^3}{6!} - 8\frac{k^5}{8!} - 8\frac{k^7}{10!} - ...\right). \quad (3.154)$$

The above derivative has only one root. Let us denote it as $k_{v\,max}$. Then,

$$\left.\frac{dJ^v_{IAE}(k)}{dk}\right|_{k=k_{v\,max}} = 0 \qquad (3.155)$$

and from equation (3.154) we obtain

$$\left.\frac{dJ^v_{IAE}(k)}{dk}\right|_{k=2} > 0 \qquad (3.156)$$

and

$$\left.\frac{dJ^v_{IAE}(k)}{dk}\right|_{k=2.5} < 0 \qquad (3.157)$$

which shows that $2 < k_{v\,max} < 2.5$. This leads to the conclusion that actually there exists such a value $k_{v\,max} \in (2, 2.5)$, that for $k < k_{v\,max}$ function $J^v_{IAE}(k)$ increases, and for $k > k_{v\,max}$, function $J^v_{IAE}(k)$ decreases with increasing k. This ends the proof.

In the case considered in this section, i.e. when the velocity constraint is taken into account, from Theorem 1 and the following relation

$$\lim_{k\to\infty} J^v_{IAE}(k) = \frac{1}{2}\frac{e_0^2}{v_{max}} < \frac{2}{e}\frac{e_0^2}{v_{max}} = \lim_{k\to 0^+} J^v_{IAE}(k), \qquad (3.158)$$

we get that the considered criterion achieves its minimum value when $k_{v\,opt} \to \infty$. Since parameter B can be found substituting $k = k_{v\,opt}$ into

$$B = \frac{v_{max}^3 k^2}{e_0^2\left(e^k-1\right)^3}\exp\left(\frac{3ke^k}{e^k-1}\right)\mathrm{sgn}(e_0) \qquad (3.159)$$

then $B_{v\,opt} \to \mathrm{sgn}(e_0)\cdot\infty$, $A_{v\,opt} \to -\,\mathrm{sgn}(e_0)\cdot\infty$, $c_{1v\,opt} \to \infty$, $c_{2v\,opt} \to \infty$ and $t_{f\,v\,opt}$ is given by relation (2.98).

Now, as we did in section 2.2.2, we will modify the criterion considered in this section so that we could introduce an elastic velocity constraint. This modified criterion is given by equation (2.99), i.e.

$$Q_{IAE}^v = \int_{t_0}^{\infty} |e_1(t)| \, dt + q \left\{ \frac{\max \left[|e_2(t)| \right]}{M} \right\}^n$$

where $M > 0$ is the threshold value of the system velocity. Taking into account the assumption that v_{max} is the threshold value of the system velocity and equation (3.37) we get

$$Q_{IAE}^v \left(k, |B| \right) = \frac{|e_0|^{4/3}}{|B|^{1/3}} \left(2 k^{-1/3} + \frac{1}{2} k^{2/3} \right) + q \left\{ \frac{\max \left[|e_2(t)| \right]}{v_{max}} \right\}^n . \tag{3.160}$$

Further in this section we minimise criterion (3.160). Since $\max[|e_2(t)|]$ is described by the absolute value of (3.149), the considered criterion can be expressed as

$$Q_{IAE}^v \left(k, |B| \right) = \frac{|e_0|^{4/3}}{|B|^{1/3}} \left(2 k^{-1/3} + \frac{1}{2} k^{2/3} \right) +$$
$$+ q \left[\exp \left(\frac{-k e^k}{e^k - 1} \right) \left(e^k - 1 \right) \frac{|B|^{1/3} e_0^{2/3}}{k^{2/3} v_{max}} \right]^n . \tag{3.161}$$

For any given k

$$\frac{\partial Q_{IAE}^v \left(k, |B| \right)}{\partial |B|} = -\frac{|e_0|^{4/3}}{3 |B|^{4/3}} \left(2 k^{-1/3} + \frac{1}{2} k^{2/3} \right) +$$
$$+ \frac{1}{3} qn |B|^{n/3-1} \left[\exp \left(\frac{-k e^k}{e^k - 1} \right) \left(e^k - 1 \right) \frac{e_0^{2/3}}{k^{2/3} v_{max}} \right]^n \tag{3.162}$$

is negative for small values of $|B|$, and becomes positive when the absolute value of B is getting bigger. As the result of solving equation $\partial Q_{IAE}^v \left(k, |B| \right) / \partial |B| = 0$ we obtain

$$|B| = \left\{ \frac{|e_0|^{4/3} \left(2 k^{-1/3} + \frac{1}{2} k^{2/3} \right)}{qn \left[\exp \left(\frac{-k e^k}{e^k - 1} \right) \left(e^k - 1 \right) \frac{e_0^{2/3}}{k^{2/3} v_{max}} \right]^n} \right\}^{\frac{3}{n+1}} . \tag{3.163}$$

3.2 Switching Plane Design Minimising IAE

Therefore, for any particular k, criterion $Q_{IAE}^v(k,|B|)$ reaches its minimum for $|B|$ expressed by (3.163). Consequently, substituting (3.163) into criterion (3.161) we get

$$Q_{IAE}^v(k,|B|) = (qn)^{\frac{1}{n+1}} \left[|e_0|^{4/3} \left(2k^{-1/3} + \frac{1}{2}k^{2/3} \right) \exp\left(\frac{-ke^k}{e^k-1}\right)(e^k-1)\frac{e_0^{2/3}}{k^{2/3}v_{max}} \right]^{\frac{n}{n+1}} +$$

$$+ q \left[\frac{1}{qn}\exp\left(\frac{-ke^k}{e^k-1}\right)(e^k-1)\frac{e_0^{2/3}|e_0|^{4/3}}{k^{2/3}v_{max}} \left(2k^{-1/3} + \frac{1}{2}k^{2/3} \right) \right]^{\frac{n}{n+1}} =$$

$$= \left[|e_0|^{4/3}\left(\frac{2}{k}+\frac{1}{2}\right)\exp\left(\frac{-ke^k}{e^k-1}\right)(e^k-1)\frac{e_0^{2/3}}{v_{max}} \right]^{\frac{n}{n+1}} \left[(qn)^{\frac{1}{n+1}} + q\left(\frac{1}{qn}\right)^{\frac{n}{n+1}} \right] =$$

$$= \left[\frac{e_0^2}{v_{max}}\left(\frac{2}{k}+\frac{1}{2}\right)\exp\left(\frac{-ke^k}{e^k-1}\right)(e^k-1) \right]^{\frac{n}{n+1}} (qn)^{\frac{1}{n+1}}\left(1+\frac{1}{n}\right).$$

$$(3.164)$$

From the above considerations and the analysis of the minimum of function (3.152), we conclude that function (3.164) achieves its minimum when k approaches infinity. Then substituting $k_{v\,opt} \to \infty$ into relation

$$B = \mathrm{sgn}(e_0)\left\{ \frac{|e_0|^{4/3}\left(2k^{-1/3} + \frac{1}{2}k^{2/3} \right)}{qn\left[\exp\left(\frac{-ke^k}{e^k-1}\right)(e^k-1)\frac{e_0^{2/3}}{k^{2/3}v_{max}} \right]^n} \right\}^{\frac{3}{n+1}}, \qquad (3.165)$$

we notice that the optimal value of parameter $|B|$ also tends to infinity. Hence, $B_{v\,opt} \to \mathrm{sgn}(e_0)\cdot\infty$, $A_{v\,opt} \to -\mathrm{sgn}(e_0)\cdot\infty$, $c_{1\,v\,opt} \to \infty$, $c_{2\,v\,opt} \to \infty$ and

$$t_{f\,v\,opt} = e_0^{1/3}\left[\frac{2nqe_0^{(2n-4)/3}}{v_{max}^n} \right]^{\frac{1}{n+1}}. \qquad (3.166)$$

It is not surprising that if $n \to \infty$, then $t_{f\,v\,opt}$ is again given by relation (2.98). This conclusion ends the analysis of the elastic velocity constraint. The plot of criterion $Q_{IAE}^v(k,|B|)$ as a function of two variables is shown in figure 3.22.

Fig. 3.22 The modified criterion $Q_{IAE}^{v}(k,|B|)$

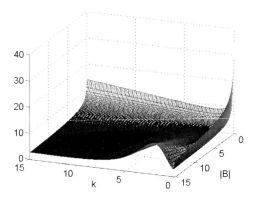

Fig. 3.23 Tracking error evolution and its first derivative (conventional velocity constraint)

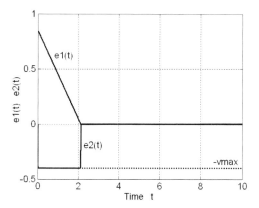

Fig. 3.24 Second order derivative of the tracking error (conventional velocity constraint)

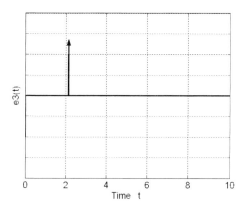

3.2 Switching Plane Design Minimising IAE

Fig. 3.25 Control signal (conventional velocity constraint)

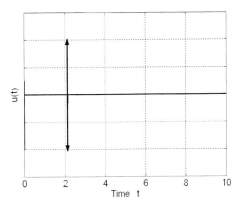

Fig. 3.26 Tracking error evolution and its first derivative ($n = 5$, $q = 5$)

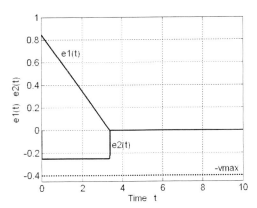

Fig. 3.27 Second order derivative of the tracking error ($n = 5$, $q = 5$)

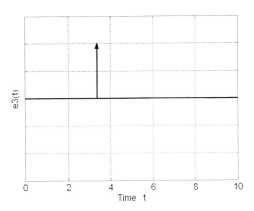

Fig. 3.28 Control signal ($n = 5$, $q = 5$)

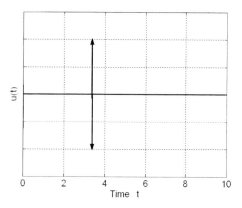

Fig. 3.29 Tracking error evolution and its first derivative ($n = 5$, $q = 0.02$)

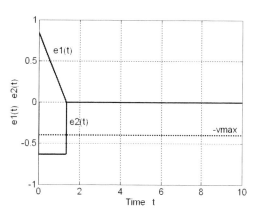

Fig. 3.30 Second order derivative of the tracking error ($n = 5$, $q = 0.02$)

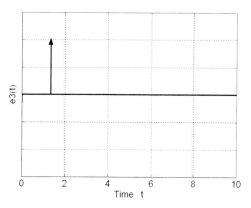

3.2 Switching Plane Design Minimising IAE

Fig. 3.31 Control signal ($n = 5$, $q = 0.02$)

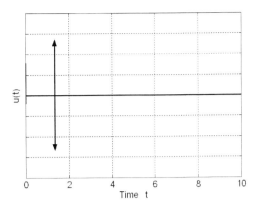

Simulation example

In order to verify the system performance subject to the velocity constraint we consider third order system (3.145) where relations (3.96) and (2.86) represent model uncertainty and external disturbance. Then, consequently $\gamma = 1$. The initial conditions and the demand trajectory are given by (3.146) and (3.147). In this example we require that the system velocity does not exceed $v_{max} = 0.4$, which implies that the plane stops moving at time instant $t_f \approx 2.125$. The tracking error evolution and its first derivative are shown in figure 3.23. It can be seen from this figure that the velocity constraint is always satisfied. The second derivative of the tracking error and the input signal are illustrated in figures 3.24 and 3.25. Figures 3.26 – 3.31 show the system performance when elastic velocity constraint is considered. In this example we take into account two cases when $n = 5$, $q = 5$ and $n = 5$, $q = 0.02$. For $n = 5$ and $q = 5$ the plane stops moving at the time $t_f \approx 3.377$ and for $n = 5$, $q = 0.02$ we obtain $t_f \approx 1.345$. It can be seen from figures 3.24, 3.25, 3.27, 3.28, 3.30 and 3.31 that the design method proposed in this section leads to infinite acceleration and requires infinite values of the control signal in the system. Therefore, it is not a feasible option for practical applications and further in this chapter we will consider the same system, but taking into account more constraints at a time.

3.2.4 Switching Plane Design Subject to Acceleration and Velocity Constraints

In this section we consider constraints (3.99) and (2.89) which are required to be satisfied at the same time. Therefore, for any k, the minimum value of $J_{IAE}^{av}(k, B)$ is given by

$$J_{IAE}^{av}(k) = \max\left[J_{IAE}^{a}(k), J_{IAE}^{v}(k) \right] \tag{3.167}$$

where criteria $J_{\text{IAE}}^{a}(k)$ and $J_{\text{IAE}}^{v}(k)$ are given by (3.109) and (3.152), respectively. Consequently, the optimal solution of the minimisation of criterion $J_{\text{IAE}}^{av}(k,B)$ is such a value of the argument k, for which

$$J_{\text{IAE}}^{av}(k_{av\,opt}) = \min_{k>0}\left\{J_{\text{IAE}}^{av}(k)\right\} = \min_{k>0}\left\{\max\left[J_{\text{IAE}}^{a}(k), J_{\text{IAE}}^{v}(k)\right]\right\} \qquad (3.168)$$

and the respective value of B. Next, we will find the optimal parameter $k_{av\,opt}$. We consider the following two cases $J_{\text{IAE}}^{v}(4) \leq J_{\text{IAE}}^{a}(4)$ and $J_{\text{IAE}}^{v}(4) > J_{\text{IAE}}^{a}(4)$. In the first case, $k_{av\,opt} = k_{a\,opt} = 4$ and the optimal value of parameter B can be calculated from (3.113). Now we will consider the second case, i.e. the situation when $J_{\text{IAE}}^{v}(4) > J_{\text{IAE}}^{a}(4)$. We begin with introducing the following theorem which is proved below.

Theorem 2

If condition $J_{\text{IAE}}^{v}(4) > J_{\text{IAE}}^{a}(4)$ is satisfied, then criterion given by expression (3.167) achieves its minimum value at a point $k_{av\,opt}$ which belongs to the open interval from 4 to $a_{max}e|e_0|/v_{max}^2$.

Proof

Notice that for any $k > 4$ criterion $J_{\text{IAE}}^{v}(k)$ is a decreasing function of its argument and for any $k > 4$ criterion $J_{\text{IAE}}^{a}(k)$ is an increasing function of k. Furthermore $\lim_{k \to \infty} J_{\text{IAE}}^{a}(k) = \infty$, so there exists such a number $k_{\alpha} \in (4, \infty)$ that $J_{\text{IAE}}^{a}(k_{\alpha}) = J_{\text{IAE}}^{v}(k_{\alpha})$. We will now demonstrate that function (3.167) achieves its minimum value at the point $k = k_{\alpha}$. For that purpose we will consider three situations, i.e. the first case when $k_{\alpha} \in (4, 6.4)$, the second when $k_{\alpha} \in [6.4, 8.5)$, and the third one when $k_{\alpha} \geq 8.5$.

i) case one $k_{\alpha} \in (4, 6.4)$

It follows from equation (3.109) that if $k_{\alpha} \leq 16$, then there exists such a number $k_{\beta} = 16/k_{\alpha} \geq 1$, that $J_{\text{IAE}}^{a}(k_{\alpha}) = J_{\text{IAE}}^{a}(k_{\beta})$. Notice that $k_{\alpha} > 4$ and consequently k_{β} is smaller than 4. Then for any $k \notin (k_{\beta}, k_{\alpha})$, $J_{\text{IAE}}^{a}(k) > J_{\text{IAE}}^{a}(k_{\alpha}) = J_{\text{IAE}}^{a}(k_{\beta})$. For any $k \in (k_{v\,max}, k_{\alpha})$ $J_{\text{IAE}}^{v}(k) > J_{\text{IAE}}^{v}(k_{\alpha}) = J_{\text{IAE}}^{a}(k_{\alpha})$ which means that if $k_{\beta} > k_{v\,max}$, then function $J_{\text{IAE}}^{av}(k)$ has its minimum at the point k_{α}. This situation takes place when $k_{\alpha} < 16/k_{v\,max}$, and taking into account inequality $k_{v\,max} < 2.5$, we conclude that if $k_{\alpha} < 16/2.5 = 6.4$, then function $J_{\text{IAE}}^{av}(k)$ must achieve its minimum value at the point k_{α}. This situation is illustrated in figure 3.32.

3.2 Switching Plane Design Minimising IAE

Fig. 3.32 Functions $J_{IAE}^{a}(k)$ and $J_{IAE}^{v}(k)$ for $k_\alpha \in (4, 6.4)$

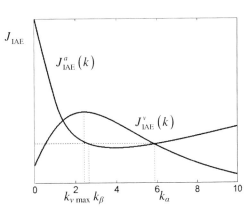

ii) case two $k_\alpha \in [6.4, 8.5)$

Now we consider the situation when $k_\alpha \in [6.4, 8.5)$, which implies that $k_\beta \in (16/8.5, 2.5]$. Function $J_{IAE}^{v}(k_\beta)$ increases for any $k_\beta \in (16/8.5, k_{v\,max})$ and decreases for any $k_\beta \in (k_{v\,max}, 2.5]$. Consequently, the minimum value of $J_{IAE}^{v}(k_\beta)$ in the considered interval $k_\beta \in (16/8.5, 2.5]$

$$\min_{k_\beta \in (16/8.5,\ 2.5]} J_{IAE}^{v}(k_\beta) = \min\left[\lim_{k_\beta \to 16/8.5} J_{IAE}^{v}(k_\beta); J_{IAE}^{v}(2.5)\right] = \tag{3.169}$$

$$= \lim_{k_\beta \to 16/8.5} J_{IAE}^{v}(k_\beta) > \frac{e_0^2}{v_{max}} 0.9447$$

is greater than the biggest value

$$\max_{k_\alpha \in [6.4,\ 8.5)} J_{IAE}^{v}(k_\alpha) = J_{IAE}^{v}(6.4) < \frac{e_0^2}{v_{max}} 0.803 \tag{3.170}$$

which implies that $J_{IAE}^{av}(k)$ achieves its minimum value at the point k_α. Figure 3.33 shows the plots of $J_{IAE}^{a}(k)$ and $J_{IAE}^{v}(k)$ in the considered case.

iii) case three $k_\alpha \geq 8.5$

Function $J_{IAE}^{v}(k)$ is continuous and decreasing for any $k > k_{v\,max}$. Furthermore, $k_{v\,max} < 4$,

$$J_{IAE}^{v}(4) > \frac{2}{e} \frac{e_0^2}{v_{max}} \tag{3.171}$$

and

$$\lim_{k \to \infty} J_{IAE}^{v}(k) = \frac{1}{2} \frac{e_0^2}{v_{max}}. \tag{3.172}$$

Fig. 3.33 Functions $J_{IAE}^a(k)$ and $J_{IAE}^v(k)$ for $k_\alpha \in [6.4, 8.5)$

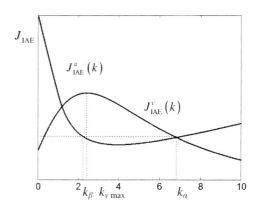

Consequently, there exists exactly one number $z \in (4, \infty)$, such that for $k \in (4, \infty)$ we have

$$J_{IAE}^v(k) < \frac{2}{e} \frac{e_0^2}{v_{max}} \Leftrightarrow k > z. \quad (3.173)$$

Moreover, from Theorem 1 and the following relation

$$J_{IAE}^v(0) = \frac{2}{e} \frac{e_0^2}{v_{max}} \quad (3.174)$$

it follows that $J_{IAE}^v(k)$ increases for any $k \in [0, k_{v\,max})$ and consequently if $k > z$, then for any $k \in [0, k_{v\,max})$, $J_{IAE}^v(k) > J_{IAE}^v(k_\alpha)$. On the other hand, $J_{IAE}^v(k)$ decreases for any $k \in (k_{v\,max}, \infty)$, so for any $k \in [k_{v\,max}, k_\alpha)$, $J_{IAE}^v(k) > J_{IAE}^v(k_\alpha)$. This implies, that if $k_\alpha > z$, then $J_{IAE}^v(k)$ achieves its minimum value at the point k_α. It is easy to verify that $z < 8.5$. Indeed

$$J_{IAE}^v(8.5) \approx 0.734 \frac{e_0^2}{v_{max}} < \frac{2}{e} \frac{e_0^2}{v_{max}}. \quad (3.175)$$

Therefore, if $k_\alpha \geq 8.5$, then function $J_{IAE}^v(k)$ has its minimum value at the point k_α. The scenario considered in this case is presented in figure 3.34.

It follows from the analysis presented above that $J_{IAE}^{av}(k)$ achieves its minimum value at a point $k_\alpha > 4$. Next, we will show that $k_\alpha < a_{max} e |e_0| / v_{max}^2$. For that purpose we define the following function

$$\tilde{J}_{IAE}^v(k) = \frac{e_0^2}{v_{max}} \left(\frac{2}{k} + \frac{1}{2} \right). \quad (3.176)$$

3.2 Switching Plane Design Minimising IAE

Fig. 3.34 Functions $J_{IAE}^a(k)$ and $J_{IAE}^v(k)$ for $k_\alpha \geq 8.5$

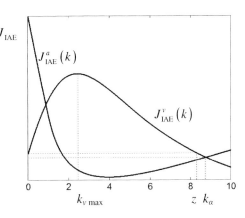

For any $k > 0$, the following relation holds

$$\tilde{J}_{IAE}^v(k) = \frac{e_0^2}{v_{max}}\left(\frac{2}{k}+\frac{1}{2}\right) > \frac{e_0^2}{v_{max}}\exp\left(\frac{-k}{e^k-1}\right)\left(\frac{2}{k}+\frac{1}{2}\right)\left(1-e^{-k}\right) =$$

$$= \frac{e_0^2}{v_{max}}\exp\left(-k-\frac{k}{e^k-1}\right)\left(\frac{2}{k}+\frac{1}{2}\right)\left(e^k-1\right) = \quad (3.177)$$

$$= \frac{e_0^2}{v_{max}}\exp\left(\frac{-ke^k}{e^k-1}\right)\left(\frac{2}{k}+\frac{1}{2}\right)\left(e^k-1\right) = J_{IAE}^v(k).$$

This implies that for any $k > 0$ function $\tilde{J}_{IAE}^v(k)$ dominates $J_{IAE}^v(k)$. Therefore, taking into account that $J_{IAE}^v(4) > J_{IAE}^a(4)$, $\lim_{k\to\infty} J_{IAE}^a(k) = \infty$ and $\lim_{k\to\infty} \tilde{J}_{IAE}^v(k) = \frac{1}{2}\frac{e_0^2}{v_{max}}$, we conclude that there exists such a number $k_\gamma > 4$, that $\tilde{J}_{IAE}^v(k_\gamma) = J_{IAE}^a(k_\gamma)$. Solving equation

$$\frac{e_0^2}{v_{max}}\left(\frac{2}{k_\gamma}+\frac{1}{2}\right) = \frac{|e_0|^{3/2}}{\sqrt{a_{max}e}}\left(\frac{2}{\sqrt{k_\gamma}}+\frac{\sqrt{k_\gamma}}{2}\right), \quad (3.178)$$

we get

$$k_\gamma = \frac{a_{max}e|e_0|}{v_{max}^2}. \quad (3.179)$$

As for any $k \in (0, \infty)$ $\tilde{J}_{IAE}^v(k) > J_{IAE}^v(k)$, it is easy to find that $k_\gamma > k_\alpha$. Thus if inequality $J_{IAE}^v(4) > J_{IAE}^a(4)$ holds, then a number $k_{av\,opt} \in (4, k_\gamma)$ is the optimal

112   3 Time-Varying Sliding Modes for the Third Order Systems

solution of criterion (3.167) minimisation task. This conclusion ends the proof of Theorem 2.

According to the presented theorem the optimal value $k_{av\,opt}$ of the parameter k belongs to the interval $(4, a_{max}e|e_0|/v_{max}^2\,)$. At this point $J_{IAE}^a\left(k_{av\,opt}\right) = J_{IAE}^v\left(k_{av\,opt}\right)$. Moreover, for any $k\in(4, a_{max}e|e_0|/v_{max}^2\,)$ criterion $J_{IAE}^a\left(k\right)$ is an increasing function of k and $J_{IAE}^v\left(k\right)$ is a decreasing function of its argument. Therefore, in order to find the optimal value $k_{av\,opt}$ we introduce the following function

$$f_3\left(k\right) = J_{IAE}^v\left(k\right) - J_{IAE}^a\left(k\right) =$$
$$= \frac{|e_0|^{3/2}}{v_{max}\sqrt{a_{max}}\,e}\left(\frac{2}{k}+\frac{1}{2}\right)\left[\sqrt{a_{max}\,e|e_0|}\exp\left(\frac{-ke^k}{e^k-1}\right)\left(e^k-1\right) - v_{max}\sqrt{k}\right]. \quad (3.180)$$

Clearly, $f_3(k)$ is monotonic in the considered interval and $f_3(4)\cdot f_3(a_{max}e|e_0|/v_{max}^2\,) < 0$. Thus $k_{av\,opt}$ which is the only root of equation $f_3(k) = 0$ in the interval can be easily found using any standard numerical procedure (for example bisection or falsi rule). The respective optimal value of B can be calculated from equations (3.112) or (3.159). Furthermore, the remaining optimal parameters of the switching plane, i.e. c_1, c_2 and A may be determined from equations (3.36), (3.15) and (3.19). The parameters determined in this way ensure the optimal performance of the controlled system together with satisfaction of both acceleration and velocity constraints.

Simulation example

In order to verify the system performance with velocity and acceleration constraints, once again we consider system (3.145) with external disturbance (2.86) and model uncertainty (3.96). Similarly as in our earlier examples also here $\gamma = 1$, initial conditions are specified by equations (3.146) and the demand trajectory is described by (3.147). The acceleration and velocity constraints are given as follows $a_{max} = 0.5$ and $v_{max} = 0.4$. According to the results presented in this section we get $c_1 \approx 3.314$, $c_2 \approx 3.641$, $B \approx 1.282$, $A \approx -2.817$ and $t_f \approx 2.197$. The tracking error evolution and its first derivative are shown in figure 3.35, and figure 3.36 shows the system acceleration. It can be seen from these figures that both constraints are satisfied and the system is insensitive with respect to external disturbance and model uncertainty from the very beginning of the control process. The input signal of the system is presented in figure 3.37.

3.2 Switching Plane Design Minimising IAE

Fig. 3.35 Tracking error evolution and its derivative

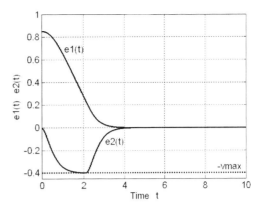

Fig. 3.36 Second derivative of the tracking error

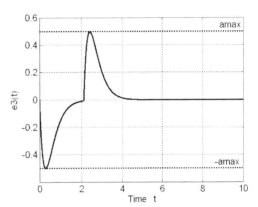

Fig. 3.37 Control signal

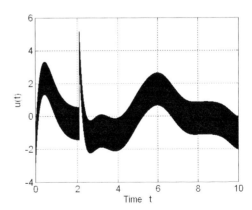

3.2.5 Switching Plane Design Subject to Acceleration and Input Signal Constraints

In this section we will consider system (3.1) subject to two constraints, i.e. now both the system acceleration and its input signal are limited. The input signal cannot be greater than u_{max} and the maximum admissible acceleration is a_{max}. Therefore, for any k, the minimum value of $J_{IAE}^{ua}(k, B)$ is given by

$$J_{IAE}^{ua}(k) = \max\left[J_{IAE}^{u}(k), J_{IAE}^{a}(k)\right] \tag{3.181}$$

where $J_{IAE}^{u}(k)$ is given by (3.51) and $J_{IAE}^{a}(k)$ by (3.109). Consequently, criterion $J_{IAE}^{ua}(k, B)$ achieves its minimum for such a value of the argument k, for which

$$J_{IAE}^{ua}(k_{ua\,opt}) = \min_{k>0}\left\{\max\left[J_{IAE}^{u}(k), J_{IAE}^{a}(k)\right]\right\} \tag{3.182}$$

and a respective value of B. Next, the optimal parameter $k_{ua\,opt}$ will be determined. In order to achieve this goal three cases:

- $J_{IAE}^{a}(k_{u\,opt}) \leq J_{IAE}^{u}(k_{u\,opt})$,
- $J_{IAE}^{a}(k_{a\,opt}) \geq J_{IAE}^{u}(k_{a\,opt})$,
- $J_{IAE}^{a}(k_{u\,opt}) > J_{IAE}^{u}(k_{u\,opt}) \wedge J_{IAE}^{a}(k_{a\,opt}) < J_{IAE}^{u}(k_{a\,opt})$

will be taken into account.

Case 1: $J_{IAE}^{a}(k_{u\,opt}) \leq J_{IAE}^{u}(k_{u\,opt})$. In this case we have

$$\min_{k>0}\left\{\max\left[J_{IAE}^{u}(k), J_{IAE}^{a}(k)\right]\right\} = J_{IAE}^{u}(k_{u\,opt}) = J_{IAE}^{u}(2). \tag{3.183}$$

This leads to the conclusion that $k_{ua\,opt} = k_{u\,opt} = 2$ and $B_{ua\,opt} = B_{u\,opt}$ can be calculated from (3.55).

Case 2: $J_{IAE}^{a}(k_{a\,opt}) \geq J_{IAE}^{u}(k_{a\,opt})$. Now we conclude that

$$\min_{k>0}\left\{\max\left[J_{IAE}^{u}(k), J_{IAE}^{a}(k)\right]\right\} = J_{IAE}^{a}(k_{a\,opt}) = J_{IAE}^{a}(4). \tag{3.184}$$

Therefore, we get $k_{ua\,opt} = k_{a\,opt} = 4$ and $B_{ua\,opt} = B_{a\,opt}$ given by (3.113).

Case 3: $J_{IAE}^{a}(k_{u\,opt}) > J_{IAE}^{u}(k_{u\,opt}) \wedge J_{IAE}^{a}(k_{a\,opt}) < J_{IAE}^{u}(k_{a\,opt})$. From the above considerations we know that for any $k \in (k_{u\,opt}, k_{a\,opt})$, $J_{IAE}^{u}(k)$ is an increasing function of k and $J_{IAE}^{a}(k)$ is a decreasing function of its argument. Taking into

3.2 Switching Plane Design Minimising IAE

account assumptions of this case, we conclude that the optimal value $k_{ua\,opt}$ of the parameter k belongs to the interval $(2, 4)$. At this point $J_{IAE}^u(k_{ua\,opt}) = J_{IAE}^a(k_{ua\,opt})$. Therefore, in order to find the optimal value $k_{ua\,opt}$ we take into account the following function

$$f_4(k) = J_{IAE}^u(k) - J_{IAE}^a(k) =$$
$$= |e_0|^{4/3}\left(2+\frac{1}{2}k\right)\left\{\frac{\left[e^{-k}(k-1)+1\right]^{1/3}}{(\delta U)^{1/3} k^{1/3}} - \frac{|e_0|^{1/6}}{\sqrt{a_{max}}\,ek}\right\}. \qquad (3.185)$$

Function $f_4(k)$ is monotonic in the considered interval and $f_4(k_{a\,opt}) \cdot f_4(k_{u\,opt}) < 0$. Thus, $k_{ua\,opt}$ – the only root of equation $f_4(k) = 0$ in the interval – can be determined using any standard numerical procedure. Then the respective optimal value $B_{ua\,opt}$ can be found from any of the relations (3.54), (3.112) and the optimal parameters c_1, c_2 and A are determined from equations (3.36), (3.15) and (3.19), respectively. The scenario considered in this optimisation task is illustrated in figure 3.38.

Fig. 3.38 Criteria $J_{IAE}^a(k)$ and $J_{IAE}^u(k)$

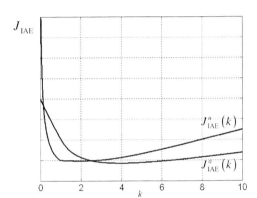

Simulation example

In this example we verify the system performance in the situation when both its acceleration and the input signal are limited. For this purpose we take into account system (3.95) with model uncertainty (3.96) and external disturbance (2.86). Consequently, just like in all the examples considered earlier, we choose $\gamma = 1$. The initial conditions are

$$x_{10} = 1.93, \quad x_{20} = 0.31, \quad x_{30} = -0.95 \qquad (3.186)$$

which results in $e_{10} = 0.98$, $e_{20} = 0$, $e_{30} = 0$ and system (3.95) is supposed to track the demand trajectory

$$x_{1d}(t) = -\sin(t - 3\pi/5) \approx -\sin(t - 1.885). \qquad (3.187)$$

Fig. 3.39 Tracking error evolution and its derivative

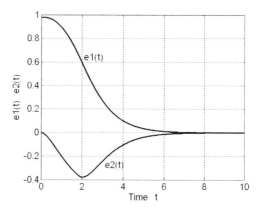

Fig. 3.40 Second derivative of the tracking error

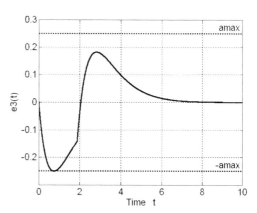

Fig. 3.41 Control signal

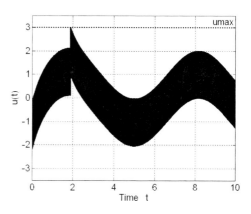

In this example we require that the maximum admissible value of the system acceleration is equal to $a_{max} = 0.25$ and the input signal is limited by $u_{max} = 3$. According to the design procedure presented in this section we obtain $c_1 \approx 1.712$,

$c_2 \approx 2.617$, $B \approx 0.889$, $A \approx -1.676$ and $t_f \approx 1.885$. The simulation results are shown in figures 3.39 – 3.41. Figure 3.39 presents the tracking error evolution and its first derivative. Then figures 3.40 and 3.41 show the two signals whose constraints are considered in this section, i.e. the acceleration and the input signal of system (3.95). One can easily notice from the figures that both constraints are always satisfied and the system is insensitive with respect to disturbance $d(t)$ and model uncertainty $\Delta f(x, t)$ right from the very beginning of the control process.

3.2.6 Switching Plane Design Subject to Input Signal and Velocity Constraints

In this section we present the switching plane design when u_{max} is the maximum admissible value of the input signal, and the velocity cannot exceed v_{max}, i.e. we take into account constraints (2.47) and (2.89). In order to find the minimum value of criterion $J_{IAE}(k, B)$ expressed by (3.37) we will minimise the following function

$$J_{IAE}^{uv}(k) = \max\left[J_{IAE}^{u}(k), J_{IAE}^{v}(k) \right]. \tag{3.188}$$

The optimal solution of the minimisation task will be such a value of k which satisfies the following condition

$$J_{IAE}^{uv}(k_{uv\,opt}) = \min_{k>0}\left\{\max\left[J_{IAE}^{u}(k), J_{IAE}^{v}(k) \right]\right\} \tag{3.189}$$

and the respective value of parameter B. First, we consider the following two cases: $J_{IAE}^{v}(2) \leq J_{IAE}^{u}(2)$ and $J_{IAE}^{v}(2) > J_{IAE}^{u}(2)$. Notice that in the case $J_{IAE}^{v}(2) \leq J_{IAE}^{u}(2)$ we get

$$\min_{k>0}\left\{\max\left[J_{IAE}^{u}(k), J_{IAE}^{v}(k) \right]\right\} = J_{IAE}^{u}(2). \tag{3.190}$$

Hence the optimal value of parameter k is $k_{uv\,opt} = k_{u\,opt} = 2$ and the optimal value $B_{uv\,opt} = B_{u\,opt}$ can be found from (3.55).

In the latter case $J_{IAE}^{v}(2) > J_{IAE}^{u}(2)$, substituting $k = 2$ into (3.152) and (3.51) after some calculations we get

$$\frac{\left|e_0\right|^{2/3}(\delta U)^{1/3}}{v_{max}} > \frac{2^{2/3}\left(e^{-2}+1\right)^{1/3}}{e^2-1}\exp\left(\frac{2e^2}{e^2-1}\right). \tag{3.191}$$

Let us denote the left hand side of this inequality by p, i.e.

$$p = \frac{\left|e_0\right|^{2/3}(\delta U)^{1/3}}{v_{max}}, \tag{3.192}$$

and its right hand side by p_1, i.e.

$$p_1 = \frac{2^{2/3}\left(e^{-2}+1\right)^{1/3}}{e^2-1}\exp\left(\frac{2e^2}{e^2-1}\right) \approx 2.619 \ . \tag{3.193}$$

Analysing evolution of $J_{IAE}^u(k)$ we notice that for $k \to 0^+$ this function increases to infinity, for $k = 2$ it reaches its minimum and when $k \to \infty$ again $J_{IAE}^u(k)$ rises to infinity. Furthermore, function $J_{IAE}^v(k)$ is an increasing function of its argument for $k \in [0, k_{v\,max})$ where $k_{v\,max} \in (2, 2.5)$ and then it decreases to $e_0^2/2v_{max}$ when k tends to infinity (compare Theorem 1). The above considerations let us conclude that if condition (3.191) is satisfied, then there exist $k = k_1 \in (0, 2)$ and $k = k_2 > 2$ which satisfy the following equations $J_{IAE}^u(k_1) = J_{IAE}^v(k_1)$ and $J_{IAE}^u(k_2) = J_{IAE}^v(k_2)$.

Notice that for $k \geq 2$ the following function

$$f_5(k) = e^{-k}(k-1)+1 \tag{3.194}$$

decreases from $f_5(2) = e^{-2}+1$ to 1 when $k \to \infty$ and is always greater than 1. This implies that, for $k \geq 2$, the following inequality is always satisfied

$$\frac{|e_0|^{4/3}}{(\delta U)^{1/3}}k^{2/3}\left(\frac{2}{k}+\frac{1}{2}\right) \leq \frac{|e_0|^{4/3}}{(\delta U)^{1/3}}\left[e^{-k}(k-1)+1\right]^{1/3}k^{2/3}\left(\frac{2}{k}+\frac{1}{2}\right). \tag{3.195}$$

In this way we determined the lower bound of criterion $J_{IAE}^u(k)$ which is equal to the right hand side of (3.195). On the other hand, since $\exp[-ke^k/(e^k-1)](e^k-1) < 1$ we have

$$\frac{e_0^2}{v_{max}}\exp\left(\frac{-ke^k}{e^k-1}\right)\left(\frac{2}{k}+\frac{1}{2}\right)\left(e^k-1\right) < \frac{e_0^2}{v_{max}}\left(\frac{2}{k}+\frac{1}{2}\right). \tag{3.196}$$

This inequality gives the upper bound of criterion $J_{IAE}^v(k)$ which matches the left hand side of (3.196). Thus solving equation

$$\frac{e_0^2}{v_{max}}\left(\frac{2}{k}+\frac{1}{2}\right) = \frac{|e_0|^{4/3}}{(\delta U)^{1/3}}k^{2/3}\left(\frac{2}{k}+\frac{1}{2}\right), \tag{3.197}$$

we obtain such a value of k that k_2 is greater than 2 and smaller than this value. Actually this value equals $p^{3/2}$, which implies that $k_2 \in (2, p^{3/2})$.

3.2 Switching Plane Design Minimising IAE

Fig. 3.42 Minimisation task of criterion $J_{IAE}^{uv}(k)$ for $p \in (p_1, p_2)$ ($p = 2.86$)

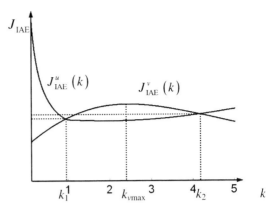

Fig. 3.43 Minimisation task of criterion $J_{IAE}^{uv}(k)$ for $p > p_2$ ($p = 4$)

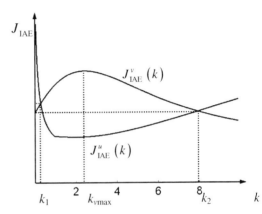

In order to find $k_{uv\,opt}$ the roots k_1 and k_2 of the following function

$$f_6(k) = J_{IAE}^{v}(k) - J_{IAE}^{u}(k) =$$

$$= \begin{cases} \dfrac{|e_0|^{4/3}}{(\delta U)^{1/3}} \left(\dfrac{2}{k}+\dfrac{1}{2}\right)\left[p(e^k-1)\exp\left(-\dfrac{ke^k}{e^k-1}\right) - k^{2/3}\right] & \text{for } k < 1, \\[2mm] \dfrac{|e_0|^{4/3}}{(\delta U)^{1/3}} \left(\dfrac{2}{k}+\dfrac{1}{2}\right)\left\{p(e^k-1)\exp\left(-\dfrac{ke^k}{e^k-1}\right) - k^{2/3}\left[e^{-k}(k-1)+1\right]^{1/3}\right\} & \text{for } k \geq 1 \end{cases}$$

(3.198)

are calculated numerically, the first one in the interval $(0, 2)$, and the second in the interval $(2, p^{3/2})$. Then we check which value $J_{IAE}^{v}(k_1) = J_{IAE}^{u}(k_1)$ or $J_{IAE}^{v}(k_2) = J_{IAE}^{u}(k_2)$ is smaller. If $J_{IAE}^{v}(k_1) = J_{IAE}^{u}(k_1)$ is smaller, then k_1 is the optimal value of k and the optimal value $B_{uv\,opt}$ of parameter B can be found from (2.58) when k_1 is smaller than 1, and can be calculated from (3.54) when k_1 is greater than or equal to 1. On the other hand, if $J^v_{\text{IAE}}\left(k_2\right) = J^u_{\text{IAE}}\left(k_2\right)$ is smaller than $J^v_{\text{IAE}}\left(k_1\right) = J^u_{\text{IAE}}\left(k_1\right)$, then k_2 is the optimal value of k and the parameter B can be found from (3.54). Furthermore, in the case when $J^v_{\text{IAE}}\left(k_1\right) = J^v_{\text{IAE}}\left(k_2\right)$ we get two optimal solutions. The first one consists of k_1 and the value of B found either from criterion (2.58) when $k_1 < 1$ or (3.54) when $k_1 \geq 1$. The second solution is given by the pair (k_2, B) where B is determined by (3.54). Moreover, no matter if $J^v_{\text{IAE}}\left(k_1\right) = J^u_{\text{IAE}}\left(k_1\right)$ is greater than $J^v_{\text{IAE}}\left(k_2\right) = J^u_{\text{IAE}}\left(k_2\right)$ or not, and for any k the respective value $B_{uv\,opt}$ can be found not only from (2.58) or (3.54) as specified above, but also from equation (3.159).

Notice that the relation between the values of $J^v_{\text{IAE}}\left(k_1\right) = J^u_{\text{IAE}}\left(k_1\right)$ and $J^v_{\text{IAE}}\left(k_2\right) = J^u_{\text{IAE}}\left(k_2\right)$ depends on p. For $p \in (p_1, p_2)$, where $p_2 \approx 3.2685$, inequality $J^v_{\text{IAE}}\left(k_1\right) < J^v_{\text{IAE}}\left(k_2\right)$ is satisfied and the optimal value of k is equal to k_1. On the other hand, for $p > p_2$ we have $J^v_{\text{IAE}}\left(k_1\right) > J^v_{\text{IAE}}\left(k_2\right)$ and the optimal solution $k = k_2$. In conclusion if $p \in (p_1, p_2)$, then the optimal value of k belongs to the interval $(0, 2)$ – this situation is shown in figure 3.42, and if $p > p_2$, then the optimal value of k belongs to the interval $(2, p^{3/2})$ – this scenario is illustrated in figure 3.43. When p is smaller than p_1, the optimal solution is always determined by the input signal constraint and $k_{uv\,opt}$ equals 2.

Simulation example

In this example, similarly as in the previous section, we consider system (3.95) with model uncertainty (3.96) and external disturbance (2.86) and we choose $\gamma = 1$. The initial conditions are specified as

$$x_{10} = 1.33, \ x_{20} = -0.71, \ x_{30} = -0.71. \tag{3.199}$$

Furthermore, system (3.95) is supposed to track the following demand trajectory

$$x_{1d}(t) = -\sin(t - \pi/4) \approx -\sin(t - 0.785). \tag{3.200}$$

The maximum admissible value of the system velocity is equal to $v_{\max} = 0.25$ and the input signal should not exceed $u_{\max} = 3$. Taking into account these constraints we find $c_1 \approx 1.25$, $c_2 \approx 2.24$, $B \approx 1$, $A \approx -0.78$ and $t_f \approx 0.785$. The tracking error evolution and its first derivative are shown in figure 3.44. It can be seen, from this figure, that the system velocity indeed does not exceed its maximum admissible value $v_{\max}$. Figure 3.45 shows the second derivative of the tracking error. Furthermore, figure 3.46 illustrates the control signal and clearly shows that the input signal constraint is also satisfied in this system. Moreover, similarly as in all the examples presented earlier in this book, the system is insensitive with respect to external disturbance and model uncertainty from the very beginning of the control action.

3.2 Switching Plane Design Minimising IAE

Fig. 3.44 Tracking error evolution and its derivative

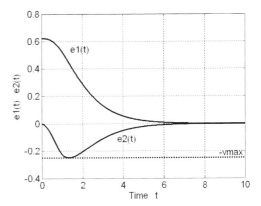

Fig. 3.45 Second derivative of the tracking error

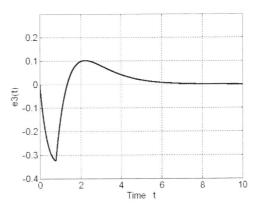

Fig. 3.46 Control signal

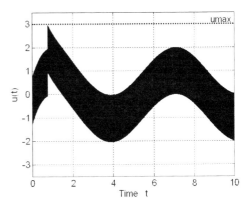

3.2.7 Switching Plane Design with Acceleration, Velocity and Input Signal Constraints

In this section we will consider system (3.1) subject to the three constraints, i.e. the system with limited acceleration, input signal and velocity. The maximum

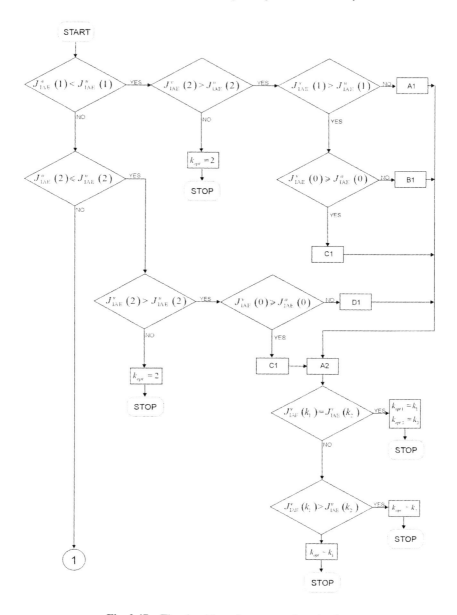

Fig. 3.47a The algorithm of parameter k_{opt} selection

3.2 Switching Plane Design Minimising IAE

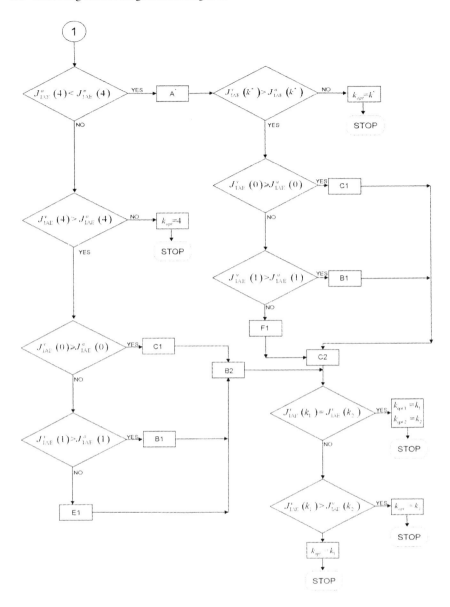

Fig. 3.47b The algorithm of parameter k_{opt} selection

admissible acceleration is a_{max}, the input signal cannot be greater than u_{max} and furthermore, the system velocity cannot exceed v_{max}. In this case, in order to find the optimal switching plane parameters, the following criterion

$$J_{IAE}^{uav}(k) = \max\left[J_{IAE}^{u}(k), J_{IAE}^{a}(k), J_{IAE}^{v}(k)\right] \tag{3.201}$$

will be minimised. Consequently, the optimal solution of the minimisation of criterion $J_{\text{IAE}}(k, B)$ with constraints (2.47), (3.99) and (2.89) is such a pair (k_{opt}, B_{opt}) that

$$J_{\text{IAE}}^{uav}\left(k_{opt}\right) = \min_{k>0}\left\{\max\left[J_{\text{IAE}}^{u}\left(k\right), J_{\text{IAE}}^{a}\left(k\right), J_{\text{IAE}}^{v}\left(k\right)\right]\right\} \tag{3.202}$$

and B_{opt} is a respective value of B. Further in this section, we will present the algorithm for finding k_{opt}. If one of the conditions i – iv given below is satisfied, then the algorithm finds k_{opt} directly. On the other hand, if any of the conditions v – xiii holds, then two values of k, i.e. k_1 and k_2, are determined and the optimal parameter k_{opt} equals either k_1 when $J_{\text{IAE}}^{v}\left(k_1\right) < J_{\text{IAE}}^{v}\left(k_2\right)$ or k_2 when $J_{\text{IAE}}^{v}\left(k_1\right) > J_{\text{IAE}}^{v}\left(k_2\right)$. In the case when $J_{\text{IAE}}^{v}\left(k_1\right) = J_{\text{IAE}}^{v}\left(k_2\right)$ we get two optimal values of k which correspond to the smallest value of the considered criterion. The block diagram of the algorithm is shown in figures 3.47a and 3.47b, and its detailed description is presented below.

i) If $J_{\text{IAE}}^{a}\left(1\right) < J_{\text{IAE}}^{u}\left(1\right) \wedge J_{\text{IAE}}^{v}\left(2\right) \leq J_{\text{IAE}}^{u}\left(2\right)$, then $k_{opt} = 2$. The optimal value B_{opt} can be calculated from formula (3.54). This is a direct consequence of the following implication

$$J_{\text{IAE}}^{a}\left(1\right) < J_{\text{IAE}}^{u}\left(1\right) \Rightarrow \bigvee_{k\geq1} J_{\text{IAE}}^{a}\left(k\right) < J_{\text{IAE}}^{u}\left(k\right). \tag{3.203}$$

In order to prove this implication we take into account inequality $J_{\text{IAE}}^{a}\left(1\right) < J_{\text{IAE}}^{u}\left(1\right)$, which is equivalent to

$$\frac{\left|e_0\right|^{3/2}}{\sqrt{a_{\max}}\,e} < \frac{\left|e_0\right|^{4/3}}{\left(\delta U\right)^{1/3}}. \tag{3.204}$$

Then calculating the difference $J_{\text{IAE}}^{u}\left(k\right) - J_{\text{IAE}}^{a}\left(k\right)$, for any $k \geq 1$ we get

$$J_{\text{IAE}}^{u}\left(k\right) - J_{\text{IAE}}^{a}\left(k\right) = \frac{\left|e_0\right|^{4/3}}{\left(\delta U\right)^{1/3}}\left[e^{-k}\left(k-1\right)+1\right]^{1/3} k^{-1/3}\left(2+\frac{k}{2}\right) +$$

$$-\frac{\left|e_0\right|^{3/2}}{\sqrt{a_{\max}}\,e}k^{-1/2}\left(2+\frac{k}{2}\right) > \frac{\left|e_0\right|^{3/2}}{\sqrt{a_{\max}}\,e}\left(2+\frac{k}{2}\right)\left\{\left[\frac{e^{-k}\left(k-1\right)+1}{k}\right]^{1/3} - \frac{1}{\sqrt{k}}\right\} \geq 0 \tag{3.205}$$

which demonstrates that implication (3.203) is true. The situation considered here is illustrated in figure 3.48.

3.2 Switching Plane Design Minimising IAE

Fig. 3.48 Criteria $J_{IAE}^{u}(k)$, $J_{IAE}^{a}(k)$ and $J_{IAE}^{v}(k)$ versus k

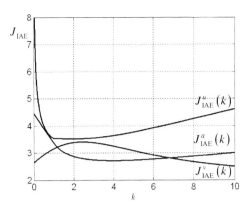

ii) If $J_{IAE}^{a}(1) \geq J_{IAE}^{u}(1) \wedge J_{IAE}^{a}(2) \leq J_{IAE}^{u}(2) \wedge J_{IAE}^{v}(2) \leq J_{IAE}^{u}(2)$, then $k_{opt} = 2$. The optimal value B_{opt} can be calculated from formula (3.54). This is justified by the following reasoning. Since $J_{IAE}^{a}(2) \leq J_{IAE}^{u}(2)$, $J_{IAE}^{v}(2) \leq J_{IAE}^{u}(2)$ and $\min_{k>0} J_{IAE}^{u}(k) = J_{IAE}^{u}(2)$ then we conclude that

$$\min_{k>0}\left\{\max\left[J_{IAE}^{u}(k), J_{IAE}^{a}(k), J_{IAE}^{v}(k)\right]\right\} = J_{IAE}^{u}(2). \qquad (3.206)$$

This is shown in figure 3.49.

iii) If $J_{IAE}^{a}(1) \geq J_{IAE}^{u}(1) \wedge J_{IAE}^{a}(2) > J_{IAE}^{u}(2) \wedge J_{IAE}^{a}(4) \geq J_{IAE}^{u}(4) \wedge \wedge J_{IAE}^{v}(4) \leq J_{IAE}^{a}(4)$, then $k_{opt} = 4$. The optimal value B_{opt} in this case is described by equation (3.112). This conclusion can be drawn directly from the fact that $J_{IAE}^{a}(k)$ reaches its minimum for $k = 4$ while both $J_{IAE}^{v}(4)$ and $J_{IAE}^{u}(4)$ are smaller than or equal to $J_{IAE}^{a}(4)$. This is illustrated in figure 3.50.

Fig. 3.49 Criteria $J_{IAE}^{u}(k)$, $J_{IAE}^{a}(k)$ and $J_{IAE}^{v}(k)$ versus k

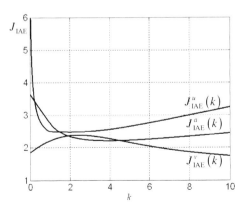

Fig. 3.50 Criteria $J_{IAE}^u(k)$, $J_{IAE}^a(k)$ and $J_{IAE}^v(k)$ versus k

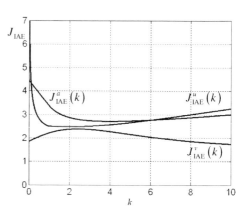

iv) If $J_{IAE}^a(1) \geq J_{IAE}^u(1) \wedge J_{IAE}^a(2) > J_{IAE}^u(2) \wedge J_{IAE}^a(4) < J_{IAE}^u(4)$, then we start with solving equation $f_4(k) = 0$ in the interval $(2, 4)$, where $f_4(k)$ is defined by equation (3.185). We denote the obtained solution as k^*. Notice that if $J_{IAE}^v(k^*) \leq J_{IAE}^a(k^*) = J_{IAE}^u(k^*)$, then $k_{opt} = k^*$. Substituting $k_{opt} = k^*$ into (3.54) or (3.112) we get the optimal value B_{opt}. In this case, which is illustrated in figure 3.51, we actually require $J_{IAE}^a(1) \geq J_{IAE}^u(1) \wedge$
$\wedge\ J_{IAE}^a(2) > J_{IAE}^u(2) \wedge J_{IAE}^a(4) < J_{IAE}^u(4) \wedge J_{IAE}^v(k^*) \leq J_{IAE}^a(k^*)$ to be satisfied. In fact the root k^* determines $\min_{k>0}\{\max[J_{IAE}^a(k), J_{IAE}^u(k)]\}$ and relation $J_{IAE}^v(k^*) \leq J_{IAE}^a(k^*) = J_{IAE}^u(k^*)$ ensures that

$$\min_{k>0}\{\max[J_{IAE}^u(k), J_{IAE}^a(k), J_{IAE}^v(k)]\} = J_{IAE}^u(k^*) = J_{IAE}^a(k^*). \quad (3.207)$$

Fig. 3.51 Criteria $J_{IAE}^u(k)$, $J_{IAE}^a(k)$ and $J_{IAE}^v(k)$ versus k

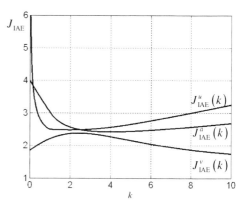

3.2 Switching Plane Design Minimising IAE

The opposite situation, i.e. $J_{IAE}^v(k^*) > J_{IAE}^a(k^*)$ is considered below under the case xiii.

Now we consider the cases which require finding two values of k and selecting the one which results in the smaller value of criterion $J_{IAE}^v(k)$. The particular values of k_1 and k_2 are determined as follows.

v) If $J_{IAE}^a(1) < J_{IAE}^u(1) \wedge J_{IAE}^v(2) > J_{IAE}^u(2) \wedge J_{IAE}^v(1) \leq J_{IAE}^u(1)$, then in order to find k_1 we solve the following equation in the interval [1, 2)

$$J_{IAE}^v(k) - J_{IAE}^u(k) =$$
$$= \frac{|e_0|^{4/3}}{(\delta U)^{1/3}} \left(\frac{2}{k} + \frac{1}{2}\right) \left\{ p(e^k - 1)\exp\left(-\frac{ke^k}{e^k - 1}\right) - k^{2/3}\left[e^{-k}(k-1) + 1\right]^{1/3} \right\} = 0.$$

(3.208)

Furthermore, solving the same equation in the interval $(2, p^{3/2})$ we obtain k_2. Having found k_1 and k_2, we choose the optimal value of k. Then substituting the optimal value k_{opt} into relations (3.54) or (3.159) we get the optimal value of parameter B. The proposed procedure for finding k_1 follows from relation (3.203) and both inequalities $J_{IAE}^v(2) > J_{IAE}^u(2)$ and $J_{IAE}^v(1) \leq J_{IAE}^u(1)$. On the other hand, the presented method of finding k_2 is justified by relation (3.203) and the first of the two inequalities. The considered scenario is shown in figure 3.52.

vi) If $J_{IAE}^a(1) < J_{IAE}^u(1) \wedge J_{IAE}^v(2) > J_{IAE}^u(2) \wedge J_{IAE}^v(1) > J_{IAE}^u(1) \wedge$
$\wedge J_{IAE}^v(0) < J_{IAE}^u(0)$, then first the following two equations

Fig. 3.52 Criteria $J_{IAE}^u(k)$, $J_{IAE}^a(k)$ and $J_{IAE}^v(k)$ versus k

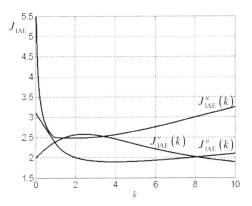

$$J_{IAE}^{v}(k) - J_{IAE}^{u}(k) =$$
$$= \frac{|e_0|^{4/3}}{(\delta U)^{1/3}} \left(\frac{2}{k} + \frac{1}{2} \right) \left[p(e^k - 1) \exp\left(-\frac{ke^k}{e^k - 1} \right) - k^{2/3} \right] = 0 \quad (3.209)$$

and

$$J_{IAE}^{v}(k) - J_{IAE}^{a}(k) =$$
$$= e_0^2 \left(2 + \frac{k}{2} \right) \left[k^{-1} v_{max}^{-1} (e^k - 1) \exp\left(-\frac{ke^k}{e^k - 1} \right) - \sqrt{\frac{1}{|e_0|a_{max} e^k}} \right] = 0 \quad (3.210)$$

are solved in the interval (0, 1). The greater of the two obtained solutions is denoted as k_1. Next, solving equation (3.208) in the interval $(2, p^{3/2})$ we get k_2 and we select the optimal value of k. Then the optimal parameter B_{opt}, for any value of k, can be found from (3.159). Notice that this is not the only way of determining B_{opt}. If the optimal value of k equals k_2, then B_{opt} can also be calculated from (3.54). Furthermore, if k_{opt} equals k_1, then B_{opt} is described by (2.58) when k_1 is the solution of equation (3.209) and B_{opt} is given by the following formula

$$B = \operatorname{sgn}(e_0) \left(\frac{a_{max}}{e^{-k} |e_0|^{1/3} k^{2/3}} \right)^{3/2} \quad (3.211)$$

when k_1 is the solution of equation (3.210). The idea of finding k_2 in this case is the same as in the case v. However, k_1 is determined differently. Conditions $J_{IAE}^{v}(1) > J_{IAE}^{u}(1)$ and $J_{IAE}^{a}(1) < J_{IAE}^{u}(1)$ imply that k_1 belongs to the interval (0, 1) and there is always one value of k for which $J_{IAE}^{u}(k) = J_{IAE}^{v}(k)$ and another one such that $J_{IAE}^{a}(k) = J_{IAE}^{v}(k)$. The

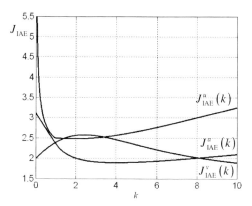

Fig. 3.53 Criteria $J_{IAE}^{u}(k)$, $J_{IAE}^{a}(k)$ and $J_{IAE}^{v}(k)$ versus k

3.2 Switching Plane Design Minimising IAE

procedure specified above determines which of the two values should be taken as k_1. Figure 3.53 presents the situation analysed here.

vii) If $J_{IAE}^{a}(1) < J_{IAE}^{u}(1) \wedge J_{IAE}^{v}(2) > J_{IAE}^{u}(2) \wedge J_{IAE}^{v}(1) > J_{IAE}^{u}(1) \wedge$
$\wedge J_{IAE}^{v}(0) \geq J_{IAE}^{a}(0)$, then we find k_1 as the solution of equation (3.209) in the interval (0, 1) and k_2 as the solution of equation (3.208) in the interval $(2, p^{3/2})$. Then we choose k_{opt}. The optimal value of B is described, as in the previous case by (3.159). Moreover, if $k_{opt} = k_1$, then B_{opt} can also be found from (2.58) and if $k_{opt} = k_2$, then B_{opt} can be determined not only from (3.159) but from (3.54) as well. The analysis of this case is similar to the previous one. The only difference is that condition $J_{IAE}^{v}(0) \geq J_{IAE}^{a}(0)$ implies that functions $J_{IAE}^{v}(k)$ and $J_{IAE}^{a}(k)$ are never equal in the interval (0, 1). Consequently, only equation (3.209) has to be solved to determine k_1. This situation is illustrated in figure 3.54.

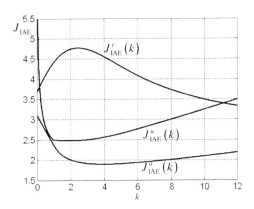

Fig. 3.54 Criteria $J_{IAE}^{u}(k)$, $J_{IAE}^{a}(k)$ and $J_{IAE}^{v}(k)$ versus k

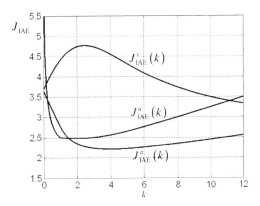

Fig. 3.55 Criteria $J_{IAE}^{u}(k)$, $J_{IAE}^{a}(k)$ and $J_{IAE}^{v}(k)$ versus k

viii) If $J^a_{IAE}(1) \geq J^u_{IAE}(1) \; \wedge \; J^a_{IAE}(2) \leq J^u_{IAE}(2) \; \wedge \; J^v_{IAE}(2) > J^u_{IAE}(2) \; \wedge$
$\wedge \; J^v_{IAE}(0) \geq J^a_{IAE}(0)$, then solving equation (3.209) in the interval (0, 1) we get k_1 and solving equation (3.208) in the interval $(2, p^{3/2})$ we obtain k_2. Having found the two values, we take one of them as k_{opt}. The way of determining parameter B and also the detailed analysis of this case are presented above, in the case vii. The illustration of this scenario is shown in figure 3.55.

ix) If $J^a_{IAE}(1) \geq J^u_{IAE}(1) \; \wedge \; J^a_{IAE}(2) \leq J^u_{IAE}(2) \; \wedge \; J^v_{IAE}(2) > J^u_{IAE}(2) \; \wedge$
$\wedge \; J^v_{IAE}(0) < J^a_{IAE}(0)$, then we solve two equations in the interval (0, 2). The first one $f_6(k) = 0$ where $f_6(k)$ is described by (3.198) and the second one $J^v_{IAE}(k) - J^a_{IAE}(k) = 0$ where

$$J^v_{IAE}(k) - J^a_{IAE}(k) =$$
$$= \begin{cases} e_0^2\left(2 + \dfrac{k}{2}\right)\left[k^{-1}v^{-1}_{\max}\left(e^k - 1\right)\exp\left(-\dfrac{ke^k}{e^k - 1}\right) - \sqrt{\dfrac{1}{|e_0|a_{\max}e^k}}\right] & \text{for } k \leq 1, \\[4mm] e_0^2\left(2 + \dfrac{k}{2}\right)\left[k^{-1}v^{-1}_{\max}\left(e^k - 1\right)\exp\left(-\dfrac{ke^k}{e^k - 1}\right) - \sqrt{\dfrac{1}{|e_0|a_{\max}ek}}\right] & \text{for } k > 1. \end{cases}$$
$$(3.212)$$

Now we denote the greater of the two obtained solutions as k_1. Moreover, solving equation (3.208) in the interval $(2, p^{3/2})$ we get k_2. Then we select k_{opt}. No matter if $k_{opt} = k_1$ or $k_{opt} = k_2$, the optimal switching plane parameter B_{opt} can be determined from (3.159). Moreover, if $k_{opt} = k_2$, then B_{opt} can also be calculated from equation (3.54). If k_{opt} is equal to k_1 and it belongs to the interval (0, 1], then B_{opt} is given by (2.58) when k_1 is the solution of equation $J^v_{IAE}(k) - J^u_{IAE}(k) = 0$ or by (3.211) when it is the solution of equation $J^v_{IAE}(k) - J^a_{IAE}(k) = 0$. On the other hand, if $k_{opt} = k_1 \in (1, 2)$, then in order to find B_{opt} we substitute k_{opt} into (3.54) when k_1 is the solution of equation $J^v_{IAE}(k) - J^u_{IAE}(k) = 0$ or into (3.112) when k_1 has been found by solving equation $J^v_{IAE}(k) - J^a_{IAE}(k) = 0$. The way of finding k_2 is the same as in the cases vii and viii. However, in order to find k_1 we solve equation $J^v_{IAE}(k) - J^a_{IAE}(k) = 0$ which is a consequence of $J^v_{IAE}(0)$ being smaller than or equal to $J^a_{IAE}(0)$. Furthermore, since inequality $J^v_{IAE}(2) > J^u_{IAE}(2)$ is satisfied, we also solve equation $J^v_{IAE}(k) - J^u_{IAE}(k) = 0$. Both of the two

3.2 Switching Plane Design Minimising IAE

Fig. 3.56 Criteria $J_{IAE}^u(k)$, $J_{IAE}^a(k)$ and $J_{IAE}^v(k)$ versus k

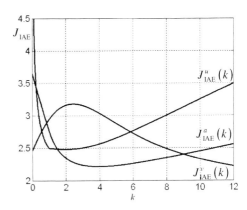

equations are solved in the interval $(0, 2)$. The example of the situation considered here is presented in figure 3.56.

x) If $J_{IAE}^a(1) \geq J_{IAE}^u(1)$ $\wedge$ $J_{IAE}^a(2) > J_{IAE}^u(2)$ $\wedge$ $J_{IAE}^a(4) \geq J_{IAE}^u(4)$ $\wedge$
$\wedge$ $J_{IAE}^v(4) > J_{IAE}^a(4)$ $\wedge$ $J_{IAE}^v(0) < J_{IAE}^a(0)$ $\wedge$ $J_{IAE}^v(1) \leq J_{IAE}^a(1)$, then
we find k_1 solving equation $f_3(k) = 0$ in the interval $[1, 4)$ where f_3 is defined by (3.180). In order to find k_2 we solve equation $f_3(k) = 0$ in the interval $(4, a_{max} ele_0/v_{max}^2)$ and equation (3.208) in the interval $(2, p^{3/2})$. Then we choose this one of the two solutions which ensures greater value of criterion $J_{IAE}^v(k)$. Once we have obtained k_1 and k_2, we check which of the two values should be adopted as k_{opt}. The optimal value of B can be calculated from (3.159). Alternatively, when $k_{opt} = k_1$ or $k_{opt} = k_2$ which is the root of equation $f_3(k) = 0$, B_{opt} can be found from (3.112) and if $k_{opt} = k_2$ determined by solving (3.208), then B_{opt} may also be determined from (3.54). Conditions $J_{IAE}^v(4) > J_{IAE}^a(4)$ and $J_{IAE}^v(1) \leq J_{IAE}^a(1)$ imply that there exists only one argument k in the interval $[1, 4)$ such that $J_{IAE}^v(k) = J_{IAE}^a(k)$. Since $J_{IAE}^a(2) > J_{IAE}^u(2)$, then criterion $J_{IAE}^u(k)$ does not play any significant role in the process of finding k_1 and in this case k_1 can be directly found solving equation $J_{IAE}^v(k) = J_{IAE}^a(k)$. In order to find k_2 equations (3.208) and $f_3(k) = 0$ are solved. This is because $J_{IAE}^a(4) \geq J_{IAE}^u(4)$, $J_{IAE}^v(4) > J_{IAE}^a(4)$, for any $k > 4$ functions $J_{IAE}^a(k)$ and $J_{IAE}^u(k)$ increase to infinity, $J_{IAE}^u(k)$ rises faster than $J_{IAE}^a(k)$ and $J_{IAE}^v(k)$ is a decreasing function of its argument. Figure 3.57 shows the relationship between $J_{IAE}^u(k)$, $J_{IAE}^a(k)$ and $J_{IAE}^v(k)$ in the considered situation.

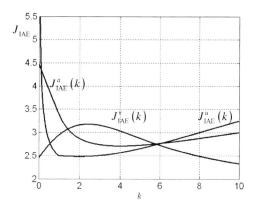

Fig. 3.57 Criteria $J_{IAE}^u(k)$, $J_{IAE}^a(k)$ and $J_{IAE}^v(k)$ versus k

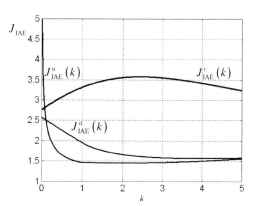

Fig. 3.58 Criteria $J_{IAE}^u(k)$, $J_{IAE}^a(k)$ and $J_{IAE}^v(k)$ versus k

xi) If $J_{IAE}^a(1) \geq J_{IAE}^u(1) \; \wedge \; J_{IAE}^a(2) > J_{IAE}^u(2) \; \wedge \; J_{IAE}^a(4) \geq J_{IAE}^u(4) \; \wedge \; \wedge \; J_{IAE}^v(4) > J_{IAE}^a(4) \; \wedge \; J_{IAE}^v(0) \geq J_{IAE}^a(0)$, then in order to find k_1 we solve equation (3.209) in the interval (0, 1). Furthermore, we solve equation $f_3(k) = 0$ in the interval $(4, a_{max}e|e_0|/v_{max}^2)$, where f_3 is given by (3.180), and equation (3.208) in the interval $(2, p^{3/2})$ and we denote by k_2 this solution which ensures the greater value of criterion $J_{IAE}^v(k)$. Then we choose k_{opt}, and in order to find the optimal parameter B_{opt} we calculate its value from (3.159). If $k_{opt} = k_1$, then B_{opt} can also be obtained from (2.58). Otherwise, i.e. if $k_{opt} = k_2$, then the following two situations are possible: either k_2 is the solution of (3.208) and then B_{opt} can also be found from (3.54) or k_2 is the root of equation $f_3(k) = 0$ and then B_{opt} can be calculated from (3.112) as well as from (3.159). Conditions $J_{IAE}^a(1) \geq J_{IAE}^u(1)$ and $J_{IAE}^v(0) > J_{IAE}^a(0)$ let us conclude that k_1 can be found from equation (3.209) in the interval (0, 1). The procedure for finding k_2 is the same as in the previous case. Figure 3.58 shows the relationship between criteria $J_{IAE}^u(k)$, $J_{IAE}^a(k)$ and $J_{IAE}^v(k)$ around k_1.

3.2 Switching Plane Design Minimising IAE   133

xii) If $J_{IAE}^a(1) \geq J_{IAE}^u(1) \wedge J_{IAE}^a(2) > J_{IAE}^u(2) \wedge J_{IAE}^a(4) \geq J_{IAE}^u(4) \wedge$
$\wedge J_{IAE}^v(4) > J_{IAE}^a(4) \wedge J_{IAE}^v(0) < J_{IAE}^a(0) \wedge J_{IAE}^v(1) > J_{IAE}^a(1)$, then equations (3.209) and (3.210) are solved in the interval (0, 1) and the greater of the two solutions is denoted by k_1. Next, in order to find k_2, equation $f_3(k) = 0$ where f_3 is given by (3.180), is solved in the interval $(4, a_{max}ele_0/v_{max}^2)$, and the root of equation (3.208) is found in the interval $(2, p^{3/2})$. In this way two values of k are obtained and this one of the two solutions which translates in the greater value of criterion $J_{IAE}^v(k)$ is denoted by k_2. After obtaining k_1 and k_2 we decide which of them is taken as k_{opt}. Then the optimal value B_{opt} can be found from (3.159). Alternatively, B_{opt} might be determined as follows. If $k_{opt} = k_1$ and k_1 is the solution of equation (3.209), then B_{opt} is given by (2.58). Furthermore, if $k_{opt} = k_1$ but k_1 is the solution of equation (3.210), then B_{opt} can be calculated from (3.211). On the other hand, if $k_{opt} = k_2$, then B_{opt} can be found from (3.112) when k_2 is the root of equation $f_3(k) = 0$ and B_{opt} may be determined from (3.54) when k_2 has been found from equation (3.208). In this case, the conditions let us calculate k_1 in the interval (0, 1). This is because in this interval $J_{IAE}^a(k)$ and $J_{IAE}^u(k)$ decrease, $J_{IAE}^v(k)$ rises and furthermore $J_{IAE}^v(0) < J_{IAE}^a(0)$, $J_{IAE}^v(1) > J_{IAE}^a(1)$ and $J_{IAE}^a(1) > J_{IAE}^u(1)$. Furthermore, k_2 is determined as described in the cases x and xi. This is illustrated in figure 3.59.

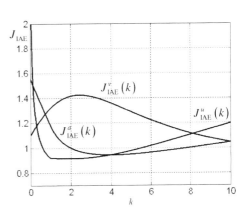

Fig. 3.59 Criteria $J_{IAE}^u(k)$, $J_{IAE}^a(k)$ and $J_{IAE}^v(k)$ versus k

xiii) If $J_{IAE}^a(1) \geq J_{IAE}^u(1) \wedge J_{IAE}^a(2) > J_{IAE}^u(2) \wedge J_{IAE}^a(4) < J_{IAE}^u(4)$, then similarly as in the case iv we find k^* solving equation $f_4(k) = 0$ in the interval (2, 4). The root k^* determines $\min_{k>0}\{\max[J_{IAE}^a(k), J_{IAE}^u(k)]\}$. Next, we consider the following three cases:

○ If $J^v_{IAE}(k^*) > J^a_{IAE}(k^*) = J^u_{IAE}(k^*) \wedge J^v_{IAE}(0) \geq J^a_{IAE}(0)$, then solving equation (3.209) in the interval (0, 1) we get k_1. Next, in order to find k_2 we solve equation (3.208) in the interval $(k^*, p^{3/2})$ and we choose k_{opt}. Furthermore, if $k_{opt} = k_1$, then B_{opt} is determined by (2.58) or (3.159) and if $k_{opt} = k_2$, B_{opt} can be found from (3.54) or (3.159). Conditions $J^a_{IAE}(1) \geq J^u_{IAE}(1)$, $J^v_{IAE}(k^*) > J^a_{IAE}(k^*) = J^u_{IAE}(k^*)$ and $J^v_{IAE}(0) \geq J^a_{IAE}(0)$ determine the way of finding k_1, and conditions $J^v_{IAE}(k^*) > J^a_{IAE}(k^*) = J^u_{IAE}(k^*)$, $J^a_{IAE}(4) < J^u_{IAE}(4)$ and the fact that $J^u_{IAE}(k)$ increases faster than $J^a_{IAE}(k)$ for any k greater than 4 determine the procedure for finding k_2. Figure 3.60 demonstrates the circumstances considered in this case.

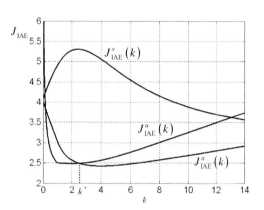

Fig. 3.60 Criteria $J^u_{IAE}(k)$, $J^a_{IAE}(k)$ and $J^v_{IAE}(k)$ versus k

○ If $J^v_{IAE}(k^*) > J^a_{IAE}(k^*) = J^u_{IAE}(k^*) \wedge J^v_{IAE}(0) < J^a_{IAE}(0) \wedge \wedge J^v_{IAE}(1) \leq J^a_{IAE}(1)$, then first, we solve equation (3.210) in the interval $[1, k^*)$ to obtain k_1. Next, we solve equation (3.208) in the interval $(k^*, p^{3/2})$ and we denote this solution by k_2. Then we select k_{opt}. If $k_{opt} = k_1$ then the optimal value B_{opt} can be calculated from (3.112) or (3.159), and otherwise i.e. when $k_{opt} = k_2$, B_{opt} is given by (3.54) or (3.159). Since $J^v_{IAE}(1) \leq J^a_{IAE}(1)$, $J^v_{IAE}(k^*) > J^a_{IAE}(k^*) = J^u_{IAE}(k^*)$ and $J^a_{IAE}(k)$ is greater than $J^u_{IAE}(k)$ in the interval $[1, k^*)$, then the value of k_1 can be found as the solution of equation (3.210) in this interval. The procedure for finding k_2 is the same as in the previous subcase. This situation is illustrated in figure 3.61.

3.2 Switching Plane Design Minimising IAE

Fig. 3.61 Criteria $J_{IAE}^{u}(k)$, $J_{IAE}^{a}(k)$ and $J_{IAE}^{v}(k)$ versus k

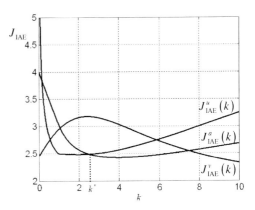

○ If $J_{IAE}^{v}(k^*) > J_{IAE}^{a}(k^*) = J_{IAE}^{u}(k^*) \wedge J_{IAE}^{v}(0) < J_{IAE}^{a}(0) \wedge$
$\wedge J_{IAE}^{v}(1) > J_{IAE}^{a}(1)$, then equations (3.209) and (3.210) are solved in the interval (0, 1) and the greater of the two obtained solutions is denoted by k_1. Moreover, k_2 is calculated solving equation (3.208) in the interval $(k^*, p^{3/2})$. Having determined k_1 and k_2, we decide which of them is taken as k_{opt}. If $k_{opt} = k_1$ found by solving equation (3.209), then B_{opt} is given by (2.58) or (3.159). Furthermore, if $k_{opt} = k_1$ calculated from equation (3.210), then B_{opt} is obtained from (3.159) or (3.211). On the other hand if $k_{opt} = k_2$, then substituting k_{opt} into (3.54) or (3.159) we get B_{opt}. Conditions $J_{IAE}^{v}(0) < J_{IAE}^{a}(0)$, $J_{IAE}^{v}(1) > J_{IAE}^{a}(1)$ and $J_{IAE}^{a}(1) \geq J_{IAE}^{u}(1)$ ensure that there exists such value of k in the interval (0, 1) that $J_{IAE}^{v}(k) = J_{IAE}^{a}(k)$ and another value of k in the same interval such that $J_{IAE}^{v}(k) = J_{IAE}^{u}(k)$. This justifies the way of determining k_1. The procedure for finding k_2 is the

Fig. 3.62 Criteria $J_{IAE}^{u}(k)$, $J_{IAE}^{a}(k)$ and $J_{IAE}^{v}(k)$ versus k

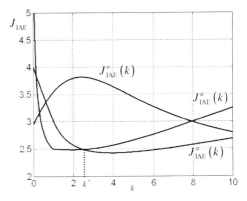

same as in the two previous subcases. The relationship between criteria $J_{IAE}^{u}(k)$, $J_{IAE}^{a}(k)$ and $J_{IAE}^{v}(k)$ in the considered situation is presented in figure 3.62.

The algorithm presented above is illustrated by the block diagram shown in figures 3.47a and 3.47b, in which blocks A1, B1, C1, D1, E1, F1 denote procedures for finding k_1, A^* denotes a procedure for finding k^* and blocks A2, B2 and C2 denote procedures for finding k_2. All these procedures are specified below.

Procedures for finding k_1

Procedure A1: Solve equation

$$\frac{|e_0|^{4/3}}{(\delta U)^{1/3}}\left(\frac{2}{k}+\frac{1}{2}\right)\left\{p\left(e^k-1\right)\exp\left(-\frac{ke^k}{e^k-1}\right)-k^{2/3}\left[e^{-k}\left(k-1\right)+1\right]^{1/3}\right\}=0$$

in the interval $k \in [1, 2)$.

Procedure B1: Solve the following two equations

$$\frac{|e_0|^{4/3}}{(\delta U)^{1/3}}\left(\frac{2}{k}+\frac{1}{2}\right)\left[p\left(e^k-1\right)\exp\left(-\frac{ke^k}{e^k-1}\right)-k^{2/3}\right]=0$$

and

$$e_0^2\left(2+\frac{k}{2}\right)\left[k^{-1}v_{max}^{-1}\left(e^k-1\right)\exp\left(-\frac{ke^k}{e^k-1}\right)-\sqrt{\frac{1}{|e_0|a_{max}e^k}}\right]=0$$

in the interval $(0, 1)$. Denote by k_1 the greater of the obtained solutions.

Procedure C1: Solve equation

$$\frac{|e_0|^{4/3}}{(\delta U)^{1/3}}\left(\frac{2}{k}+\frac{1}{2}\right)\left[p\left(e^k-1\right)\exp\left(-\frac{ke^k}{e^k-1}\right)-k^{2/3}\right]=0$$

in the interval $(0, 1)$.

Procedure D1: Solve the following two equations

$$0=\begin{cases}\dfrac{|e_0|^{4/3}}{(\delta U)^{1/3}}\left(\dfrac{2}{k}+\dfrac{1}{2}\right)\left[p\left(e^k-1\right)\exp\left(-\dfrac{ke^k}{e^k-1}\right)-k^{2/3}\right] & \text{for } k<1 \\[3ex] \dfrac{|e_0|^{4/3}}{(\delta U)^{1/3}}\left(\dfrac{2}{k}+\dfrac{1}{2}\right)\cdot \\[2ex] \cdot\left\{p\left(e^k-1\right)\exp\left(-\dfrac{ke^k}{e^k-1}\right)-k^{2/3}\left[e^{-k}\left(k-1\right)+1\right]^{1/3}\right\} & \text{for } k\geq1\end{cases}$$

3.2 Switching Plane Design Minimising IAE

and

$$0 = \begin{cases} e_0^2\left(2+\dfrac{k}{2}\right)\left[k^{-1}v_{max}^{-1}\left(e^k-1\right)\exp\left(-\dfrac{ke^k}{e^k-1}\right)-\sqrt{\dfrac{1}{|e_0|a_{max}e^k}}\right] & \text{for } k \le 1 \\[4mm] e_0^2\left(2+\dfrac{k}{2}\right)\left[k^{-1}v_{max}^{-1}\left(e^k-1\right)\exp\left(-\dfrac{ke^k}{e^k-1}\right)-\sqrt{\dfrac{1}{|e_0|a_{max}ek}}\right] & \text{for } k > 1 \end{cases}$$

in the interval $(0, 2)$. Denote by k_1 the greater of the two obtained solutions.

Procedure E1: Solve equation

$$\frac{|e_0|^{3/2}}{v_{max}\sqrt{a_{max}}e}\left(\frac{2}{k}+\frac{1}{2}\right)\left[\sqrt{a_{max}e|e_0|}\exp\left(\frac{-ke^k}{e^k-1}\right)\left(e^k-1\right)-v_{max}\sqrt{k}\right]=0$$

in the interval $[1, 4)$.

Procedure F1: Solve equation

$$\frac{|e_0|^{3/2}}{v_{max}\sqrt{a_{max}}e}\left(\frac{2}{k}+\frac{1}{2}\right)\left[\sqrt{a_{max}e|e_0|}\exp\left(\frac{-ke^k}{e^k-1}\right)\left(e^k-1\right)-v_{max}\sqrt{k}\right]=0$$

in the interval $[1, k^*)$.

Procedure for finding k^*

Procedure A*: Solve equation

$$|e_0|^{4/3}\left(2+\frac{1}{2}k\right)\left\{\frac{\left[e^{-k}(k-1)+1\right]^{1/3}}{(\delta U)^{1/3}k^{1/3}}-\frac{|e_0|^{1/6}}{\sqrt{a_{max}}ek}\right\}=0$$

in the interval $(2, 4)$.

Procedures for finding k_2

Procedure A2: Solve equation

$$\frac{|e_0|^{4/3}}{(\delta U)^{1/3}}\left(\frac{2}{k}+\frac{1}{2}\right)\left\{p\left(e^k-1\right)\exp\left(-\frac{ke^k}{e^k-1}\right)-k^{2/3}\left[e^{-k}(k-1)+1\right]^{1/3}\right\}=0$$

in the interval $(2, p^{3/2})$.

Procedure B2: Solve equation

$$\frac{|e_0|^{3/2}}{v_{max}\sqrt{a_{max}}e}\left(\frac{2}{k}+\frac{1}{2}\right)\left[\sqrt{a_{max}e|e_0|}\exp\left(\frac{-ke^k}{e^k-1}\right)\left(e^k-1\right)-v_{max}\sqrt{k}\right]=0$$

in the interval $(4, a_{max}e|e_0|/v_{max}^2)$, and equation

$$\frac{|e_0|^{4/3}}{(\delta U)^{1/3}}\left(\frac{2}{k}+\frac{1}{2}\right)\left\{p\left(e^k-1\right)\exp\left(-\frac{ke^k}{e^k-1}\right)-k^{2/3}\left[e^{-k}(k-1)+1\right]^{1/3}\right\}=0$$

in the interval $(2, p^{3/2})$. Denote by k_2 this one of the two obtained solutions which ensures the greater value of criterion $J_{IAE}^v(k)$.

Procedure C2: Solve equation

$$\frac{|e_0|^{4/3}}{(\delta U)^{1/3}}\left(\frac{2}{k}+\frac{1}{2}\right)\left\{p\left(e^k-1\right)\exp\left(-\frac{ke^k}{e^k-1}\right)-k^{2/3}\left[e^{-k}(k-1)+1\right]^{1/3}\right\}=0$$

in the interval $(k^*, p^{3/2})$.

The algorithm presented in this section generates the optimal switching plane parameters k and B for system (3.1) subject to all the three constraints considered in the book. The other parameters of the plane, i.e. c_1, c_2, A and t_f can be found from equations (3.36), (3.15), (3.19) and (3.23), respectively. That concludes the methods for finding switching plane parameters which ensure the minimisation of the integral absolute error (IAE) with the input signal, acceleration and velocity constraint.

Simulation example

In order to verify the system performance when the input signal is required to be satisfied we consider third order system (3.95) where relations (3.96) and (2.86) represent model uncertainty and external disturbance. Then, consequently $\gamma = 1$. The initial conditions are as follows

$$x_{10} = 0.8, \ x_{20} = 1, \ x_{30} = 0. \quad (3.213)$$

System (3.95) is supposed to track the demand trajectory

$$x_{1d}(t) = \sin t. \quad (3.214)$$

In this case we require that the maximum admissible values of the system velocity, acceleration and input signal are equal to $v_{max} = 0.25$, $a_{max} = 0.18$ and $u_{max} = 3$, respectively. According to the presented method we obtain $c_1 \approx 3.71$, $c_2 \approx 3.95$, $B \approx 0.94$, $A \approx -2.96$ and $t_f \approx 3.14$. The tracking error evolution and its first derivative are presented in figure 3.63. Figure 3.64 shows the second derivative of the tracking error and figure 3.65 demonstrates the control signal. It can be seen from

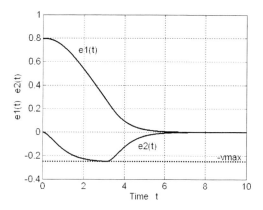

Fig. 3.63 Tracking error evolution and its first derivative

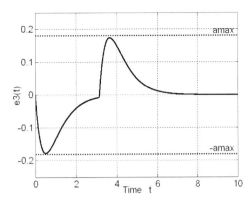

Fig. 3.64 Second derivative of the tracking error

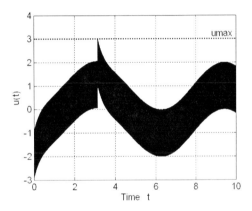

Fig. 3.65 Control signal

the three figures that all the constraints are satisfied at the same time. Furthermore, the system is insensitive with respect to external disturbance and model uncertainty from the very beginning of the control action.

That ends the description of the methods for switching plane design which minimise IAE. In the next section new algorithms ensuring ITAE minimisation will be presented.

3.3 Switching Plane Design Minimising ITAE

In the remaining part of this monograph we still consider third order system (3.1) with velocity, acceleration and input signal constraints. However, in order to select the parameters of the switching plane (3.7), (3.8) now we minimise the integral of the time multiplied by the absolute error (ITAE). For this purpose, first we calculate the ITAE defined by (2.120). Because we have demonstrated in section 3.1 that the tracking error converges monotonically in the considered system, then we may replace criterion (2.120) with (2.121). Next, substituting (3.20) and (3.28) into (2.121) and calculating appropriate integrals, we get

$$
J_{\text{ITAE}} = \int_0^{\frac{e_0 c_1}{B}} \left[t \left(-\frac{2B\sqrt{c_1}}{c_1^2} - \frac{B}{c_1} t \right) e^{-\sqrt{c_1} t} + t \frac{2B\sqrt{c_1}}{c_1^2} + t e_0 - \frac{B}{c_1} t^2 \right] dt +
$$

$$
+ \int_{\frac{e_0 c_1}{B}}^{\infty} t e^{-\sqrt{c_1} t} \left[-\frac{2B\sqrt{c_1}}{c_1^2} + \frac{2B\sqrt{c_1}}{c_1^2} e^k - e_0 e^k + \left(-\frac{B}{c_1} + \frac{B}{c_1} e^k \right) t \right] dt = \qquad (3.215)
$$

$$
= \frac{e_0^2 \sqrt{c_1}}{|B|} + \frac{|e_0|^3 c_1^2}{6B^2} + \frac{3|e_0|}{c_1}.
$$

Since the considered constraints, i.e. input signal, acceleration and velocity constraint, are expressed in terms of k rather than c_1, further in this monograph it will be convenient to consider quality criterion $J_{\text{ITAE}}(k, B)$ instead of $J_{\text{ITAE}}(c_1, B)$. Taking into account relation (3.24) we can present the criterion as

$$
J_{\text{ITAE}} = \frac{|e_0|^{5/3}}{|B|^{2/3}} \left(k^{1/3} + \frac{1}{6} k^{4/3} + 3k^{-2/3} \right). \qquad (3.216)
$$

Next in this section, criterion (3.216) will be minimised, in order to determine the switching plane parameters.

3.3.1 Switching Plane Design Subject to Input Signal Constraint

First we will precisely analyse the minimisation of criterion (3.216) subject to input signal constraint (2.47). Notice that for any given value of k, the minimum of criterion (3.216) is obtained for the greatest value of $|B|$ satisfying constraints (2.55) for $k < 1$ and (3.50) for $k \geq 1$. Therefore, the solution of the minimisation task can be found as a minimum of a single variable function. Consequently, in order to minimise criterion (3.216) with constraints (2.55) and (3.50) the following two cases $k < 1$ and $k \geq 1$, should be considered. Thus, we obtain the following performance index expressed as a function of a single variable

$$
J_{\text{ITAE}}^u (k) = \begin{cases} \dfrac{|e_0|^{5/3}}{(\delta U)^{2/3}} \left(k^{1/3} + \dfrac{1}{6} k^{4/3} + 3k^{-2/3} \right) & \text{for } k < 1, \\[4mm] \dfrac{|e_0|^{5/3}}{(\delta U)^{2/3}} \left[e^{-k} (k-1) + 1 \right]^{2/3} \left(k^{1/3} + \dfrac{1}{6} k^{4/3} + 3k^{-2/3} \right) & \text{for } k \geq 1. \end{cases} \qquad (3.217)
$$

3.3 Switching Plane Design Minimising ITAE

This function is shown in figure 3.66.

Fig. 3.66 Criterion $J^u_{ITAE}(k)$

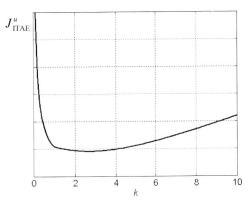

In order to solve the minimisation task we calculate the derivative of expression (3.217)

$$\frac{dJ^u_{ITAE}(k)}{dk} = \begin{cases} \dfrac{2|e_0|^{5/3}}{3(\delta U)^{2/3}} k^{-5/3}\left(\dfrac{1}{3}k^2 + \dfrac{1}{2}k - 3\right) & \text{for } k < 1, \\ \dfrac{2|e_0|^{5/3}}{3(\delta U)^{2/3}}\left[e^{-k}(k-1)+1\right]^{-1/3} k^{-5/3} e^{-k}\left[e^k\left(\dfrac{1}{3}k^2+\dfrac{1}{2}k-3\right) + \\ \quad -\dfrac{1}{6}k^4 - \dfrac{1}{3}k^3 - \dfrac{5}{6}k^2 + \dfrac{5}{2}k + 3\right] & \text{for } k \geq 1. \end{cases} \quad (3.218)$$

Notice that when $k < 1$ the above derivative is always negative. Hence, for $k \in (0, 1)$ function J^u_{ITAE} has its minimum value at the point $k \to 1$. This minimum value equals

$$J^u_{ITAE}(1) = \frac{25|e_0|^{5/3}}{6(\delta U)^{2/3}} \approx 4.167\frac{|e_0|^{5/3}}{(\delta U)^{2/3}}. \quad (3.219)$$

In the second case, i.e. when $k \geq 1$ we introduce the following function

$$g(k) = e^k\left(\frac{1}{3}k^2 + \frac{1}{2}k - 3\right) - \frac{1}{6}k^4 - \frac{1}{3}k^3 - \frac{5}{6}k^2 + \frac{5}{2}k + 3. \quad (3.220)$$

From (3.220) and Euler's equation we have

$$g(k) = \left(1 + k + \frac{1}{2}k^2 + \frac{1}{6}k^3 + \frac{1}{24}k^4 + \ldots\right)\left(\frac{1}{3}k^2 + \frac{1}{2}k - 3\right) +$$
$$-\frac{1}{6}k^4 - \frac{1}{3}k^3 - \frac{5}{6}k^2 + \frac{5}{2}k + 3 = \ldots \tag{3.221}$$
$$= -\frac{3}{2}k^2 - \frac{1}{4}k^3 - \frac{1}{24}k^4 + \frac{37}{30 \cdot 24}k^5 + \frac{1}{3 \cdot 24}k^6 + \ldots$$

This leads to the conclusion that for $k > 0$ function $g(k)$ has exactly one root, and at this point with the increase of the argument k, the sign of the derivative $dJ^u_{ITAE}(k)/dk$ changes from negative to positive. Furthermore, $g(2) < 0$ and $g(3) > 0$. Consequently, it can be calculated numerically that for $k = k_0 \approx 2.706$ we have $g(k) = 0$. Thus, $J^u_{ITAE}(k)$ has a single minimum, which is equal to

$$J^u_{ITAE}(k_0) = \frac{|e_0|^{5/3}}{(\delta U)^{2/3}}\left[e^{-k_0}(k_0 - 1) + 1\right]^{2/3}\left(k_0^{1/3} + \frac{1}{6}k_0^{4/3} + 3k_0^{-2/3}\right) \approx$$
$$\approx 3.833 \cdot \frac{|e_0|^{5/3}}{(\delta U)^{2/3}} \tag{3.222}$$

for $k = k_0$. Notice that expression (3.222) is smaller than (3.219). Hence, it can be concluded that the following switching plane parameters are the optimal solution of the considered minimisation problem

$$k_{u\,opt} = k_0 \approx 2.706, \tag{3.223}$$

$$B_{u\,opt} = \delta U\,\text{sgn}(e_0)\big/\left[e^{-k_{u\,opt}}(k_{u\,opt} - 1) + 1\right] \approx 0.898 \cdot \delta U\,\text{sgn}(e_0). \tag{3.224}$$

Consequently, from equations (3.36), (3.15) and (3.19) we get the other parameters of the switching plane

$$c_{1u\,opt} = \left\{\frac{k_{u\,opt}\delta U}{|e_0|\left[e^{-k_{u\,opt}}(k_{u\,opt} - 1) + 1\right]}\right\}^{\frac{2}{3}} \approx 1.807 \cdot \left(\frac{\delta U}{|e_0|}\right)^{\frac{2}{3}}, \tag{3.225}$$

$$c_{2u\,opt} = 2\left\{\frac{k_{u\,opt}\delta U}{|e_0|\left[e^{-k_{u\,opt}}(k_{u\,opt} - 1) + 1\right]}\right\}^{\frac{1}{3}} \approx 2.688 \cdot \left(\frac{\delta U}{|e_0|}\right)^{\frac{1}{3}}, \tag{3.226}$$

$$A_{u\,opt} = -e_0\left\{\frac{k_{u\,opt}\delta U}{|e_0|\left[e^{-k_{u\,opt}}(k_{u\,opt} - 1) + 1\right]}\right\}^{\frac{2}{3}} \approx -1.807 \cdot e_0\left(\frac{\delta U}{|e_0|}\right)^{\frac{2}{3}}. \tag{3.227}$$

3.3 Switching Plane Design Minimising ITAE

The switching plane described by these parameters stops moving at the time instant

$$t_{f\,u\,opt} = k_{u\,opt}^{2/3} \left\{ \frac{\left| e_0 \right| \left[e^{-k_{u\,opt}} \left(k_{u\,opt} - 1 \right) + 1 \right]}{\delta U} \right\}^{\frac{1}{3}} \approx 2.013 \cdot \left(\frac{\left| e_0 \right|}{\delta U} \right)^{\frac{1}{3}}. \tag{3.228}$$

Now we move on to the analysis of the system subject to elastic constraint. We take into account the modified criterion which, in this case, may be presented as

$$Q_{ITAE}^u = \int_{t_0}^{\infty} t \left| e_1(t) \right| dt + q \left\{ \frac{\max \left[\left| u(t) \right| \right]}{M} \right\}^n \tag{3.229}$$

where (like in section 3.2.1) $M > 0$ is the threshold value of the input signal, $q > 0$ is a weighting factor and $n \geq 1$ is a constant determining how elastic the constraint is. Since the tracking error converges monotonically in the considered system, criterion (3.229) is equivalent to

$$Q_{ITAE}^u = \left| \int_0^{\infty} t e_1(t) dt \right| + q \left\{ \frac{\max \left[\left| u(t) \right| \right]}{M} \right\}^n. \tag{3.230}$$

Taking into account equation (3.216) and inequality (3.42), similarly as in section 3.2.1, we get

$$Q_{ITAE}^u = \frac{\left| e_0 \right|^{5/3}}{\left| B \right|^{2/3}} \left(k^{1/3} + \frac{1}{6} k^{4/3} + 3 k^{-2/3} \right) + q \left\{ \frac{\max \left[\left| \dot{e}_3(t) \right| \right]}{\delta U} \right\}^n. \tag{3.231}$$

Now we will thoroughly analyse the minimisation task. We will consider the following two situations (as we did it in section 3.2.1): the first when $k < 1$ and the second one when $k \geq 1$. We begin with the case when $k < 1$. In this situation we obtain

$$Q_{ITAE}^u \left(k, \left| B \right| \right) = \frac{\left| e_0 \right|^{5/3}}{\left| B \right|^{2/3}} \left(k^{1/3} + \frac{1}{6} k^{4/3} + 3 k^{-2/3} \right) + q \left(\frac{\left| B \right|}{\delta U} \right)^n. \tag{3.232}$$

This criterion is a function of two variables k and $|B|$ and it can be minimised by finding its partial derivatives with respect to both variables and equating these derivatives to zero. The partial derivatives have the following form

$$\frac{\partial Q_{ITAE}^u \left(k, \left| B \right| \right)}{\partial k} = \frac{\left| e_0 \right|^{5/3}}{\left| B \right|^{2/3}} \left(\frac{1}{3} k^{-2/3} + \frac{2}{9} k^{1/3} - 2 k^{-5/3} \right), \tag{3.233}$$

$$\frac{\partial Q_{\text{ITAE}}^{u}\left(k,|B|\right)}{\partial |B|} = -\frac{2\left|e_0\right|^{5/3}}{3|B|^{5/3}}\left(k^{1/3} + \frac{1}{6}k^{4/3} + 3k^{-2/3}\right) + qn\frac{|B|^{n-1}}{\left(\delta U\right)^{n}}. \tag{3.234}$$

Since for any $k < 1$ the right hand side of (3.233) is negative, then criterion (3.232) decreases with increasing k. Hence we analyse the situation when $k \to 1$. Furthermore, if $\partial Q_{\text{ITAE}}^{u}\left(k,|B|\right)/\partial |B|$ equals zero, then we have

$$|B|^{n+\frac{2}{3}} = \frac{2\left|e_0\right|^{5/3}\left(\delta U\right)^{n}}{3qn}\left(k^{1/3} + \frac{1}{6}k^{4/3} + 3k^{-2/3}\right) \tag{3.235}$$

and putting $k \to 1$, we get

$$|B_1| = \left[\frac{25\left|e_0\right|^{5/3}\left(\delta U\right)^{n}}{9qn}\right]^{\frac{3}{3n+2}}. \tag{3.236}$$

In the considered case, criterion (3.232) has the following value

$$Q_{\text{ITAE}}^{u}\left(1,|B_1|\right) = \frac{25\left|e_0\right|^{5/3}\left(9qn\right)^{\frac{2}{3n+2}}}{6\left[25\left|e_0\right|^{5/3}\left(\delta U\right)^{n}\right]^{\frac{2}{3n+2}}} + \frac{q}{\left(\delta U\right)^{n}}\left[\frac{25\left|e_0\right|^{5/3}\left(\delta U\right)^{n}}{9qn}\right]^{\frac{3n}{3n+2}}. \tag{3.237}$$

In the other case, i.e. when $k \geq 1$ we minimise expression

$$Q_{\text{ITAE}}^{u}\left(k,|B|\right) = \frac{\left|e_0\right|^{5/3}}{|B|^{2/3}}\left(k^{1/3} + \frac{1}{6}k^{4/3} + 3k^{-2/3}\right) + q\left\{\frac{|B|\left[e^{-k}\left(k-1\right)+1\right]}{\delta U}\right\}^{n}. \tag{3.238}$$

In order to find the extreme value of this expression, we calculate its partial derivatives. These derivatives are given by the following formulae

$$\frac{\partial Q_{\text{ITAE}}^{u}\left(k,|B|\right)}{\partial k} = \frac{\left|e_0\right|^{5/3}}{|B|^{2/3}}\left(\frac{1}{3}k^{-2/3} + \frac{2}{9}k^{1/3} - 2k^{-5/3}\right) +$$
$$+ qn\left(\frac{|B|}{\delta U}\right)^{n}\left[e^{-k}\left(k-1\right)+1\right]^{n-1}e^{-k}\left(2-k\right), \tag{3.239}$$

$$\frac{\partial Q_{\text{ITAE}}^{u}\left(k,|B|\right)}{\partial |B|} = -\frac{2\left|e_0\right|^{5/3}}{3|B|^{5/3}}\left(k^{1/3} + \frac{1}{6}k^{4/3} + 3k^{-2/3}\right) +$$
$$+ qn\frac{|B|^{n-1}}{\left(\delta U\right)^{n}}\left[e^{-k}\left(k-1\right)+1\right]^{n}. \tag{3.240}$$

3.3 Switching Plane Design Minimising ITAE

Then solving equation $\partial Q_{\text{ITAE}}^u \left(k, |B| \right) / \partial |B| = 0$, we get

$$|B| = \left\{ \frac{2 |e_0|^{5/3} (\delta U)^n}{3qn \left[e^{-k} (k-1)+1 \right]^n} \left(k^{1/3} + \frac{1}{6} k^{4/3} + 3k^{-2/3} \right) \right\}^{\frac{3}{3n+2}}. \tag{3.241}$$

Furthermore, from relation (3.241) and equation $\partial Q_{\text{ITAE}}^u \left(k, |B| \right) / \partial k = 0$ we obtain

$$\left(k-1+e^k \right) \left(\frac{1}{2} k + \frac{1}{3} k^2 - 3 \right) + \left(2-k \right) \left(k^2 + \frac{1}{6} k^3 + 3k \right) = 0. \tag{3.242}$$

Hence, after some calculations we get

$$e^k \left(\frac{1}{3} k^2 + \frac{1}{2} k - 3 \right) - \frac{1}{6} k^4 - \frac{1}{3} k^3 - \frac{5}{6} k^2 + \frac{5}{2} k + 3 = 0. \tag{3.243}$$

Notice that the left hand side of equation (3.243) is equal to the function $g(k)$ defined by (3.220). Consequently, repeating our previous considerations, we get the solution $k = k_0 \approx 2.706$ and we expect criterion (3.238) to have its minimum value at $k = k_0$ and the following value of $|B|$

$$|B_2| = \left\{ \frac{2 |e_0|^{5/3} (\delta U)^n \left(k_0^{1/3} + \frac{1}{6} k_0^{4/3} + 3k_0^{-2/3} \right)}{3qn \left[e^{-k_0} (k_0-1)+1 \right]^n} \right\}^{\frac{3}{3n+2}} \approx$$

$$\approx \left[\frac{2.378 \cdot |e_0|^{5/3} (\delta U)^n}{qn (1.114)^n} \right]^{\frac{3}{3n+2}}. \tag{3.244}$$

In order to verify this, we calculate the second order partial derivatives

$$\frac{\partial^2 Q_{\text{ITAE}}^u \left(k, |B| \right)}{\partial k \partial k} = \frac{|e_0|^{5/3}}{|B|^{2/3}} \left(-\frac{2}{9} k^{-5/3} + \frac{2}{27} k^{-2/3} + \frac{10}{3} k^{-8/3} \right) +$$

$$+ qn \left(\frac{|B|}{\delta U} \right)^n \left\{ (n-1) \left[e^{-k} (k-1)+1 \right]^{n-2} \left[e^{-k} (2-k) \right]^2 + \left[e^{-k} (k-1)+1 \right]^{n-1} e^{-k} (k-3) \right\}$$

$$\tag{3.245}$$

$$\frac{\partial^2 Q_{\mathrm{ITAE}}^u\left(k,|B|\right)}{\partial k\partial|B|}=\frac{\partial^2 Q_{\mathrm{ITAE}}^u\left(k,|B|\right)}{\partial|B|\partial k}=$$

$$=-\frac{2|e_0|^{5/3}}{3|B|^{5/3}}\left(\frac{1}{3}k^{-2/3}+\frac{2}{9}k^{1/3}-2k^{-5/3}\right)+ \tag{3.246}$$

$$+qn^2\frac{|B|^{n-1}}{(\delta U)^n}\left[e^{-k}\left(k-1\right)+1\right]^{n-1}e^{-k}\left(2-k\right),$$

$$\frac{\partial^2 Q_{\mathrm{ITAE}}^u\left(k,|B|\right)}{\partial|B|\partial|B|}=\frac{10|e_0|^{5/3}}{9|B|^{8/3}}\left(k^{1/3}+\frac{1}{6}k^{4/3}+3k^{-2/3}\right)+ \tag{3.247}$$

$$+qn\left(n-1\right)\frac{|B|^{n-2}}{(\delta U)^n}\left[e^{-k}\left(k-1\right)+1\right]^n.$$

Then for $k = k_0$ and $|B| = |B_2|$ we construct the following matrix

$$\mathbf{H}=\begin{bmatrix}\left.\dfrac{\partial^2 Q_{\mathrm{ITAE}}^u\left(k,|B|\right)}{\partial k\partial k}\right|_{\substack{k=k_0\\|B|=|B_2|}} & \left.\dfrac{\partial^2 Q_{\mathrm{ITAE}}^u\left(k,|B|\right)}{\partial k\partial|B|}\right|_{\substack{k=k_0\\|B|=|B_2|}}\\[4mm] \left.\dfrac{\partial^2 Q_{\mathrm{ITAE}}^u\left(k,|B|\right)}{\partial|B|\partial k}\right|_{\substack{k=k_0\\|B|=|B_2|}} & \left.\dfrac{\partial^2 Q_{\mathrm{ITAE}}^u\left(k,|B|\right)}{\partial|B|\partial|B|}\right|_{\substack{k=k_0\\|B|=|B_2|}}\end{bmatrix}. \tag{3.248}$$

Straightforward calculations, omitted for the sake of clarity, lead to the conclusion that for any $n \geq 1$, $q > 0$, $\delta > 0$ and $U > 0$ we have

$$\left.\frac{\partial^2 Q_{\mathrm{ITAE}}^u\left(k,|B|\right)}{\partial k\partial k}\right|_{\substack{k=k_0\\|B|=|B_2|}}>0 \tag{3.249}$$

and that $\det \mathbf{H} > 0$. These conditions show that matrix $\mathbf{H}$ is positive definite. Hence, we conclude that for $k = k_0$ and

$$B_2 = \mathrm{sgn}\left(e_0\right)\left\{\frac{2|e_0|^{5/3}\left(\delta U\right)^n\left(k_0^{1/3}+\dfrac{1}{6}k_0^{4/3}+3k_0^{-2/3}\right)}{3qn\left[e^{-k_0}\left(k_0-1\right)+1\right]^n}\right\}^{\frac{3}{3n+2}}\approx \tag{3.250}$$

$$\approx \mathrm{sgn}\left(e_0\right)\left[\frac{2.378\cdot|e_0|^{5/3}\left(\delta U\right)^n}{qn\left(1.114\right)^n}\right]^{\frac{3}{3n+2}},$$

the control quality criterion has its minimum value

3.3 Switching Plane Design Minimising ITAE

$$Q_{ITAE}^{u}\left(k_0,|B_2|\right) = \frac{\left(k_0^{1/3}+\frac{1}{6}k_0^{4/3}+3k_0^{-2/3}\right)|e_0|^{5/3}\left\{3qn\left[e^{-k_0}\left(k_0-1\right)+1\right]^n\right\}^{\frac{2}{3n+2}}}{\left[2\left(k_0^{1/3}+\frac{1}{6}k_0^{4/3}+3k_0^{-2/3}\right)|e_0|^{5/3}\left(\delta U\right)^n\right]^{\frac{2}{3n+2}}}+$$

$$+\frac{q\left[e^{-k_0}\left(k_0-1\right)+1\right]^n}{\left(\delta U\right)^n}\left\{\frac{2\left(k_0^{1/3}+\frac{1}{6}k_0^{4/3}+3k_0^{-2/3}\right)|e_0|^{5/3}\left(\delta U\right)^n}{3qn\left[e^{-k_0}\left(k_0-1\right)+1\right]^n}\right\}^{\frac{3n}{3n+2}} \approx \tag{3.251}$$

$$\approx \frac{3.567\cdot|e_0|^{5/3}\left[qn\left(1.114\right)^n\right]^{\frac{2}{3n+2}}}{\left[2.378\cdot|e_0|^{5/3}\left(\delta U\right)^n\right]^{\frac{2}{3n+2}}}+\frac{q\left(1.114\right)^n}{\left(\delta U\right)^n}\left[\frac{2.378\cdot|e_0|^{5/3}\left(\delta U\right)^n}{qn\left(1.114\right)^n}\right]^{\frac{3n}{3n+2}}.$$

Notice that

$$Q_{ITAE}^{u}\left(1,|B_1|\right)-Q_{ITAE}^{u}\left(k_0,|B_2|\right)=$$

$$=\left\{\left(\frac{25}{6}\right)^{\frac{3n}{3n+2}}-\left[e^{-k_0}\left(k_0-1\right)+1\right]^{\frac{2n}{3n+2}}\left(k_0^{1/3}+\frac{1}{6}k_0^{4/3}+3k_0^{-2/3}\right)^{\frac{3n}{3n+2}}\right\}\cdot \tag{3.252}$$

$$\cdot\left\{\frac{|e_0|^{5/3}\left(3qn\right)^{2/(3n+2)}}{\left[2|e_0|^{5/3}\left(\delta U\right)^n\right]^{2/(3n+2)}}+\frac{q}{\left(\delta U\right)^n}\left[\frac{|e_0|^{5/3}\left(\delta U\right)^n}{qn}\right]^{3n/(3n+2)}\left(\frac{2}{3}\right)^{3n/(3n+2)}\right\}$$

and after some calculations, we get

$$Q_{ITAE}^{u}\left(1,|B_1|\right)-Q_{ITAE}^{u}\left(k_0,|B_2|\right)=$$

$$=|e_0|^{5n/(3n+2)}\left(\delta U\right)^{-2n/(3n+2)}\cdot$$

$$\cdot\left\{\left(\frac{25}{6}\right)^{3n/(3n+2)}-\left[e^{-k_0}\left(k_0-1\right)+1\right]^{2n/(3n+2)}\left(k_0^{1/3}+\frac{1}{6}k_0^{4/3}+3k_0^{-2/3}\right)^{3n/(3n+2)}\right\}\cdot$$

$$\cdot\left[\left(\frac{3}{2}qn\right)^{2/(3n+2)}+q\left(qn\right)^{-3n/(3n+2)}\left(\frac{2}{3}\right)^{3n/(3n+2)}\right]\approx \tag{3.253}$$

$$\approx|e_0|^{5n/(3n+2)}\left(\delta U\right)^{-2n/(3n+2)}\left[\left(4.167\right)^{3n/(3n+2)}-\left(3.833\right)^{3n/(3n+2)}\right].$$

$$\cdot\left[\left(\frac{3}{2}qn\right)^{2/(3n+2)}+q\left(qn\right)^{-3n/(3n+2)}\left(\frac{2}{3}\right)^{3n/(3n+2)}\right].$$

Since q, n, U and δ are positive, we have $Q^u_{\mathrm{ITAE}}\left(1,|B_1|\right) - Q^u_{\mathrm{ITAE}}\left(k_0,|B_2|\right) > 0$. Consequently, $Q^u_{\mathrm{ITAE}}\left(k_0,|B_2|\right)$ is the minimum value of the considered criterion in both cases ($k < 1$ and $k \geq 1$). Therefore, we conclude that parameters $k_{u\,opt} = k_0 \approx 2.706$ and $B_{u\,opt} = B_2$ given by relation (3.250) are the optimal switching plane parameters, in the sense of the considered criterion, i.e. ITAE with an elastic input signal constraint. The other three parameters of the switching plane can be calculated from equations (3.36), (3.15) and (3.19)

$$
c_{1u\,opt} = \left\{ \frac{k_{u\,opt}}{|e_0|} \left[\frac{2|e_0|^{5/3}\left(\delta U\right)^n \left(k_{u\,opt}^{1/3} + \frac{1}{6}k_{u\,opt}^{4/3} + 3k_{u\,opt}^{-2/3}\right)}{3qn\left[e^{-k_{u\,opt}}\left(k_{u\,opt}-1\right)+1\right]^n} \right]^{\frac{3}{3n+2}} \right\}^{\frac{2}{3}} \approx
$$

$$
\approx 1.942 \cdot |e_0|^{\frac{-2n+2}{3n+2}} \left[\frac{2.378 \cdot \left(\delta U\right)^n}{qn\left(1.114\right)^n} \right]^{\frac{2}{3n+2}},
\tag{3.254}
$$

$$
c_{2u\,opt} = 2\left\{ \frac{k_{u\,opt}}{|e_0|} \left[\frac{2|e_0|^{5/3}\left(\delta U\right)^n \left(k_{u\,opt}^{1/3} + \frac{1}{6}k_{u\,opt}^{4/3} + 3k_{u\,opt}^{-2/3}\right)}{3qn\left[e^{-k_{u\,opt}}\left(k_{u\,opt}-1\right)+1\right]^n} \right]^{\frac{3}{3n+2}} \right\}^{\frac{1}{3}} \approx
$$

$$
\approx 2.787 \cdot |e_0|^{\frac{-n+1}{3n+2}} \left[\frac{2.378 \cdot \left(\delta U\right)^n}{qn\left(1.114\right)^n} \right]^{\frac{1}{3n+2}},
\tag{3.255}
$$

$$
A_{u\,opt} = -e_0\left\{ \frac{k_{u\,opt}}{|e_0|} \left[\frac{2|e_0|^{5/3}\left(\delta U\right)^n \left(k_{u\,opt}^{1/3} + \frac{1}{6}k_{u\,opt}^{4/3} + 3k_{u\,opt}^{-2/3}\right)}{3qn\left[e^{-k_{u\,opt}}\left(k_{u\,opt}-1\right)+1\right]^n} \right]^{\frac{3}{3n+2}} \right\}^{\frac{2}{3}} \approx
$$

$$
\approx -1.942 \cdot \mathrm{sgn}\left(e_0\right)|e_0|^{\frac{n+4}{3n+2}} \left[\frac{2.378 \cdot \left(\delta U\right)^n}{qn\left(1.114\right)^n} \right]^{\frac{2}{3n+2}}.
\tag{3.256}
$$

The switching plane described by these parameters, arrives at the origin of the error state space and stops moving at the time instant

3.3 Switching Plane Design Minimising ITAE

$$t_{f\,u\,opt} = k_{u\,opt}^{2/3} |e_0|^{1/3} \left\{ \frac{3qn\left[e^{-k_{u\,opt}}\left(k_{u\,opt}-1\right)+1\right]^n}{2|e_0|^{5/3}(\delta U)^n \left(k_{u\,opt}^{1/3} + \frac{1}{6}k_{u\,opt}^{4/3} + 3k_{u\,opt}^{-2/3}\right)} \right\}^{\frac{1}{3n+2}} \approx$$

(3.257)

$$\approx 1.942 \cdot |e_0|^{\frac{n-1}{3n+2}} \left[\frac{qn(1.114)^n}{2.378 \cdot (\delta U)^n}\right]^{\frac{1}{3n+2}}.$$

Criterion $Q_{\text{ITAE}}^u(k,|B|)$ is illustrated in figure 3.67.

Fig. 3.67 The modified criterion $Q_{\text{ITAE}}^u(k,|B|)$

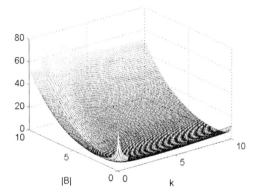

The ITAE optimal solutions determined in this part of the monograph can be easily verified in simulation examples. However, since the results of the simulations are always quite similar, and sometimes exactly the same as those already demonstrated in section 3.2, we will not present them again. We just point out that all the solutions described in various parts of section 3.3 ensure asymptotic error convergence without overshoots or oscillations and guarantee that appropriate constraints are always satisfied.

3.3.2 Switching Plane Design Subject to Acceleration Constraint

In this section we continue to select the time-varying switching planes optimal in the sense of the integral of the time multiplied by the absolute error, however now (similarly as in section 3.2.2) we take into account the system acceleration constraint. Thus, we will minimise criterion (3.216) subject to constraint (3.99). Since criterion (3.216) decreases with the increasing absolute value of B, the minimisation of the two variable function $J_{\text{ITAE}}(k,B)$ with the considered constraint may be accomplished minimising a single variable function without constraints. The maximum admissible value of |B| is derived either from constraint (3.105) or (3.108). With these constraints, criterion (3.216) can be expressed as

$$J_{ITAE}^{a}(k) = \begin{cases} \dfrac{e_0^2}{a_{max} e^k}\left(k + \dfrac{1}{6}k^2 + 3\right) & \text{for } k \leq 1 \\ \dfrac{e_0^2}{a_{max} e}\left(1 + \dfrac{1}{6}k + 3k^{-1}\right) & \text{for } k > 1 \end{cases} \quad (3.258)$$

and it is illustrated in figure 3.68.

Fig. 3.68 Criterion $J_{ITAE}^{a}(k)$

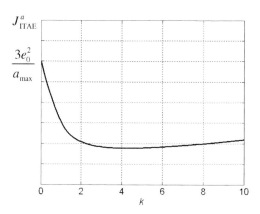

Furthermore, the analysis of criterion $J_{ITAE}^{a}(k)$ leads to the conclusion that it decreases for $k \in (0, 3\sqrt{2})$ and it increases for any $k > 3\sqrt{2}$. Therefore, for $k = k_{a\,opt} = 3\sqrt{2}$ this function reaches its minimum

$$J_{ITAE}^{a}(k_{a\,opt}) = J_{ITAE}^{a}(3\sqrt{2}) = \dfrac{e_0^2}{a_{max} e}(1 + \sqrt{2}). \quad (3.259)$$

For $k = k_{a\,opt} = 3\sqrt{2}$ we get the optimal parameter B from the following formula

$$B_{a\,opt} = \text{sgn}(e_0)\left(\dfrac{a_{max} e k_{a\,opt}^{1/3}}{|e_0|^{1/3}}\right)^{3/2} = \text{sgn}(e_0)\left[\dfrac{(3\sqrt{2})^{1/3} a_{max} e}{|e_0|^{1/3}}\right]^{3/2}. \quad (3.260)$$

The other parameters of the optimal switching plane can be found from (3.36), (3.15) and (3.19)

3.3 Switching Plane Design Minimising ITAE

$$c_{1\,a\,opt} = \frac{3\sqrt{2}a_{\max}e}{|e_0|} \tag{3.261}$$

$$c_{2\,a\,opt} = 2\sqrt{\frac{3\sqrt{2}a_{\max}e}{|e_0|}} \tag{3.262}$$

$$A_{a\,opt} = -3\sqrt{2}a_{\max}e\,\mathrm{sgn}(e_0). \tag{3.263}$$

The plane stops moving at the time instant

$$t_{f\,a\,opt} = \sqrt{\frac{3\sqrt{2}\,|e_0|}{a_{\max}e}}. \tag{3.264}$$

Now let us analyse the elastic acceleration constraint. In this case the modified criterion has the following form

$$Q^a_{\mathrm{ITAE}} = \int_{t_0}^{\infty} t\,|e_1(t)|\,dt + q\left\{\frac{\max\big[|e_3(t)|\big]}{M}\right\}^n \tag{3.265}$$

where as in section 3.2.2, $M > 0$ and $q > 0$ denote the maximum admissible value of the system acceleration and a weighting factor, respectively, and $n \geq 1$ shows how elastic the constraint is. Taking into account equation (3.216) and the assumption that $a_{\max}$ is the maximum admissible value of the system acceleration, criterion (3.265) can be written as

$$Q^a_{\mathrm{ITAE}} = \frac{|e_0|^{5/3}}{|B|^{2/3}}\left(k^{1/3} + \frac{1}{6}k^{4/3} + 3k^{-2/3}\right) + q\left\{\frac{\max\big[|e_3(t)|\big]}{a_{\max}}\right\}^n. \tag{3.266}$$

Now we will analyse the minimisation task. We will consider the following two situations: the first one which takes place when $k \leq 1$ and the second one which arises when $k > 1$. We begin with the case $k \leq 1$. In this situation, from relations (3.108) and (3.266) we have

$$Q^a_{\mathrm{ITAE}}(k,|B|) = \frac{|e_0|^{5/3}}{|B|^{2/3}}\left(k^{1/3} + \frac{1}{6}k^{4/3} + 3k^{-2/3}\right) + q\left(\frac{|e_0|^{1/3}|B|^{2/3}k^{2/3}}{e^k a_{\max}}\right)^n. \tag{3.267}$$

This criterion is a function of two independent variables k and $|B|$ and it can be minimised by finding its partial derivatives with respect to both variables

$$\frac{\partial Q^a_{ITAE}(k,|B|)}{\partial k} = \frac{|e_0|^{5/3}}{|B|^{2/3}}\left(\frac{1}{3}k^{-2/3} + \frac{2}{9}k^{1/3} - 2k^{-5/3}\right) +$$
$$+ qn\left(\frac{|e_0|^{1/3}|B|^{2/3}k^{2/3}}{e^k a_{max}}\right)^n\left(\frac{2}{3}k^{-1} - 1\right), \tag{3.268}$$

$$\frac{\partial Q^a_{ITAE}(k,|B|)}{\partial |B|} = -\frac{2|e_0|^{5/3}}{3|B|^{5/3}}\left(k^{1/3} + \frac{1}{6}k^{4/3} + 3k^{-2/3}\right) +$$
$$+ \frac{2}{3}qn|B|^{2n/3-1}\left(\frac{|e_0|^{1/3}k^{2/3}}{e^k a_{max}}\right)^n \tag{3.269}$$

and equating these derivatives to zero. Solving equation $\partial Q^a_{ITAE}(k,|B|)/\partial |B| = 0$ the following formula can be obtained

$$|B| = \left[|e_0|^{5/3}\left(\frac{e^k a_{max}}{|e_0|^{1/3}k^{2/3}}\right)^n \frac{k^{1/3} + k^{4/3}/6 + 3k^{-2/3}}{qn}\right]^{\frac{3}{2n+2}}. \tag{3.270}$$

Furthermore, equation $\partial Q^a_{ITAE}(k,|B|)/\partial k = 0$ with $|B|$ determined from relation (3.270) does not have any solution for $k \leq 1$. This leads to the conclusion that criterion (3.267) decreases with increasing k. Then putting $k = 1$ into expression (3.270) we get

$$|B_1| = \left[\frac{25|e_0|^{(5-n)/3}(ea_{max})^n}{6qn}\right]^{\frac{3}{2n+2}}. \tag{3.271}$$

In this case criterion (3.267) has the following value

$$Q^a_{ITAE}(1,|B_1|) = \frac{25}{6}|e_0|^{5/3}\left(1 + \frac{1}{n}\right)\left[\frac{25|e_0|^{(5-n)/3}(a_{max}e)^n}{6qn}\right]^{\frac{-1}{n+1}}. \tag{3.272}$$

In the second case, i.e. when $k > 1$, taking into account inequality (3.105), we minimise the following expression

$$Q^a_{ITAE} = \frac{|e_0|^{5/3}}{|B|^{2/3}}\left(k^{1/3} + \frac{1}{6}k^{4/3} + 3k^{-2/3}\right) + q\left(\frac{|e_0|^{1/3}|B|^{2/3}}{ea_{max}k^{1/3}}\right)^n. \tag{3.273}$$

Calculating partial derivatives of Q^a_{ITAE}

3.3 Switching Plane Design Minimising ITAE

$$\frac{\partial Q^a_{\mathrm{ITAE}}\left(k,|B|\right)}{\partial k} = \frac{|e_0|^{5/3}}{|B|^{2/3}}\left(\frac{1}{3}k^{-2/3} + \frac{2}{9}k^{1/3} - 2k^{-5/3}\right) - \frac{1}{3}qn\left(\frac{|e_0|^{1/3}|B|^{2/3}}{ea_{\max}k^{1/3}}\right)^n k^{-1}, \quad (3.274)$$

$$\frac{\partial Q^a_{\mathrm{ITAE}}\left(k,|B|\right)}{\partial |B|} = -\frac{2|e_0|^{5/3}}{3|B|^{5/3}}\left(k^{1/3} + \frac{1}{6}k^{4/3} + 3k^{-2/3}\right) + \frac{2}{3}qn|B|^{2n/3-1}\left(\frac{|e_0|^{1/3}}{ea_{\max}k^{1/3}}\right)^n \quad (3.275)$$

and solving equation $\partial Q^a_{\mathrm{ITAE}}/\partial|B| = 0$, we get

$$|B| = \left[\frac{|e_0|^{(5-n)/3}}{qn}\left(ea_{\max}k^{1/3}\right)^n\left(k^{1/3} + \frac{1}{6}k^{4/3} + 3k^{-2/3}\right)\right]^{\frac{3}{2n+2}}. \quad (3.276)$$

Furthermore, from relation (3.276) and equation $\partial Q^a_{\mathrm{ITAE}}/\partial k = 0$ we obtain another equation

$$\left(\frac{|e_0|}{k}\right)^{5/3}\left[\frac{|e_0|^{(5-n)/3}}{qn}\left(ea_{\max}k^{1/3}\right)^n\left(k^{1/3} + \frac{1}{6}k^{4/3} + 3k^{-2/3}\right)\right]^{\frac{-2}{2n+2}}\left(\frac{1}{6}k^2 - 3\right) = 0 \quad (3.277)$$

whose solution is equal to $k_0 = 3\sqrt{2}$ and we expect criterion (3.273) to reach its minimum value at this k_0 and the following value of $|B|$

$$|B_2| = \left[3^{1/3}\frac{|e_0|^{(5-n)/3}}{qn}\left(3\sqrt{2}\right)^{n/3}\left(ea_{\max}\right)^n\left(2^{2/3} + 2^{1/6}\right)\right]^{\frac{3}{2n+2}}. \quad (3.278)$$

It can be verified that for $k = k_0 = 3\sqrt{2}$ and $|B| = |B_2|$ the matrix of the second order partial derivatives of Q^a_{ITAE} is positive definite. In order to build the matrix we calculate the second order partial derivatives

$$\frac{\partial^2 Q^a_{\mathrm{ITAE}}\left(k,|B|\right)}{\partial k \partial k} = \frac{|e_0|^{5/3}}{|B|^{2/3}}k^{-8/3}\left(-\frac{2}{9}k + \frac{2}{27}k^2 + \frac{10}{3}\right) +$$
$$+ \frac{1}{3}\left(\frac{n}{3}+1\right)qn\left(\frac{|e_0|^{1/3}|B|^{2/3}}{ea_{\max}k^{1/3}}\right)^n k^{-2}, \quad (3.279)$$

$$\frac{\partial^2 Q^a_{\mathrm{ITAE}}\left(k,|B|\right)}{\partial k \partial |B|} = \frac{\partial^2 Q^a_{\mathrm{ITAE}}\left(k,|B|\right)}{\partial |B| \partial k} =$$
$$= -\frac{2|e_0|^{5/3}}{3|B|^{5/3}}k^{-5/3}\left(\frac{1}{3}k + \frac{2}{9}k^2 - 2\right) - \frac{2}{9}qn^2|B|^{2n/3-1}\left(\frac{|e_0|^{1/3}}{ea_{\max}k^{1/3}}\right)^n k^{-1}, \quad (3.280)$$

$$\frac{\partial^2 Q^a_{ITAE}(k,|B|)}{\partial|B|\partial|B|} = \frac{10|e_0|^{5/3}}{9|B|^{8/3}} k^{-2/3}\left(k + \frac{1}{6}k^2 + 3\right) +$$

$$+ \frac{2}{3}\left(\frac{2}{3}n - 1\right)qn|B|^{2n/3-2}\left(\frac{|e_0|^{1/3}}{ea_{max}\,k^{1/3}}\right)^n. \tag{3.281}$$

Then for $k = k_0 = 3\sqrt{2}$ and $|B| = |B_2|$ we construct the following matrix

$$\boldsymbol{H} = \begin{bmatrix} h_{11} & h_{12} \\ h_{21} & h_{22} \end{bmatrix} = \begin{bmatrix} \left.\dfrac{\partial^2 Q^a_{ITAE}(k,|B|)}{\partial k \partial k}\right|_{\substack{k=k_0 \\ |B|=|B_2|}} & \left.\dfrac{\partial^2 Q^a_{ITAE}(k,|B|)}{\partial k \partial|B|}\right|_{\substack{k=k_0 \\ |B|=|B_2|}} \\[4mm] \left.\dfrac{\partial^2 Q^a_{ITAE}(k,|B|)}{\partial|B| \partial k}\right|_{\substack{k=k_0 \\ |B|=|B_2|}} & \left.\dfrac{\partial^2 Q^a_{ITAE}(k,|B|)}{\partial|B| \partial|B|}\right|_{\substack{k=k_0 \\ |B|=|B_2|}} \end{bmatrix} \tag{3.282}$$

where

$$h_{11} = \frac{|e_0|^{5/3}\left(3\sqrt{2}\right)^{-8/3}(qn)^{\frac{2}{2n+2}}\left[\left(2+\sqrt{2}\right)(n+1)+18\right]}{3\left[3^{1/3}|e_0|^{(5-n)/3}\left(3\sqrt{2}\right)^{n/3}(ea_{max})^n\left(2^{2/3}+2^{1/6}\right)\right]^{\frac{2}{2n+2}}}, \tag{3.283}$$

$$h_{12} = h_{21} = \frac{-2|e_0|^{5/3}\left(3\sqrt{2}\right)^{-5/3}(qn)^{\frac{5}{2n+2}}\left(2+\sqrt{2}\right)(n+1)}{3\left[3^{1/3}|e_0|^{(5-n)/3}\left(3\sqrt{2}\right)^{n/3}(ea_{max})^n\left(2^{2/3}+2^{1/6}\right)\right]^{\frac{5}{2n+2}}}, \tag{3.284}$$

$$h_{22} = \frac{4|e_0|^{5/3}(qn)^{\frac{8}{2n+2}}\left(3\sqrt{2}\right)^{-2/3}\left(2+\sqrt{2}\right)(n+1)}{3\left[3^{1/3}|e_0|^{(5-n)/3}\left(3\sqrt{2}\right)^{n/3}(ea_{max})^n\left(2^{2/3}+2^{1/6}\right)\right]^{\frac{8}{2n+2}}}. \tag{3.285}$$

Notice that

$$\frac{|e_0|^{5/3}\left(3\sqrt{2}\right)^{-8/3}(qn)^{\frac{2}{2n+2}}\left[n\left(2+\sqrt{2}\right)+\sqrt{2}+20\right]}{3\left[3^{1/3}|e_0|^{(5-n)/3}\left(3\sqrt{2}\right)^{n/3}(ea_{max})^n\left(2^{2/3}+2^{1/6}\right)\right]^{\frac{2}{2n+2}}} > 0 \tag{3.286}$$

and

3.3 Switching Plane Design Minimising ITAE

$$\det \boldsymbol{H} = \frac{8|e_0|^{10/3}\,(qn)^{\frac{10}{2n+2}}\left(2+\sqrt{2}\right)(n+1)}{\left(3\sqrt{2}\right)^{10/3}\left[3^{1/3}\,|e_0|^{(5-n)/3}\left(3\sqrt{2}\right)^{n/3}\left(ea_{\max}\right)^n\left(2^{2/3}+2^{1/6}\right)\right]^{\frac{10}{2n+2}}} > 0 \qquad (3.287)$$

which shows that matrix $\boldsymbol{H}$ is indeed positive definite. Hence, we conclude that for $k_0 = 3\sqrt{2}$ and

$$B_2 = \left[3^{1/3}\,\frac{|e_0|^{(5-n)/3}}{qn}\left(3\sqrt{2}\right)^{n/3}\left(ea_{\max}\right)^n\left(2^{2/3}+2^{1/6}\right)\right]^{\frac{3}{2n+2}}\operatorname{sgn}\left(e_0\right), \qquad (3.288)$$

control quality criterion (3.273) reaches its minimum value

$$Q^a_{\text{ITAE}}\left(3\sqrt{2},B_2\right) = \left(2^{2/3}+2^{1/6}\right)|e_0|^{5/3}\left(1+\frac{1}{n}\right)\left(\sqrt{2}\right)^{\frac{-n}{3(n+1)}}\cdot$$
$$\cdot\left[\frac{|e_0|^{(5-n)/3}}{qn}\left(ea_{\max}\right)^n\left(2^{2/3}+2^{1/6}\right)\right]^{\frac{-1}{n+1}}. \qquad (3.289)$$

Notice that since q, n and $a_{\max}$ are positive, we have

$$Q^a_{\text{ITAE}}\left(1,|B_1|\right) - Q^a_{\text{ITAE}}\left(3\sqrt{2},|B_2|\right) =$$
$$= |e_0|^{5/3}\left(1+\frac{1}{n}\right)\left[|e_0|^{(5-n)/3}\,\frac{\left(a_{\max}e\right)^n}{qn}\right]^{\frac{-1}{n+1}}\left[\left(\frac{25}{6}\right)^{\frac{n}{n+1}}-\left(\sqrt{2}+1\right)^{\frac{n}{n+1}}\right] > 0. \qquad (3.290)$$

Consequently, $Q^a_{\text{ITAE}}\left(3\sqrt{2},|B_2|\right)$ is the smallest possible value of the considered criterion in both cases ($k \leq 1$ and $k > 1$). This leads to the conclusion that parameters $k_{a\,opt} = 3\sqrt{2}$ and $B_{a\,opt} = B_2$ described by relation (3.288) are the optimal switching plane parameters, in the sense of the considered criterion. The other parameters of the switching plane can be found from equations

$$c_{1a\,opt} = \left(3\sqrt{2}/|e_0|\right)^{2/3}\left[\frac{|e_0|^{(5-n)/3}}{qn}\left(3\sqrt{2}\right)^{n/3}3^{1/3}\left(ea_{\max}\right)^n\left(2^{2/3}+2^{1/6}\right)\right]^{\frac{1}{n+1}}, \qquad (3.291)$$

$$c_{2a\,opt} = 2\left(3\sqrt{2}/|e_0|\right)^{1/3}\left[\frac{|e_0|^{(5-n)/3}}{qn}\left(3\sqrt{2}\right)^{n/3}3^{1/3}\left(ea_{\max}\right)^n\left(2^{2/3}+2^{1/6}\right)\right]^{\frac{1}{2(n+1)}}, \qquad (3.292)$$

$$A_{a\,opt} = -e_0\left(3\sqrt{2}/|e_0|\right)^{2/3}\left[\frac{|e_0|^{(5-n)/3}}{qn}\left(3\sqrt{2}\right)^{n/3} 3^{1/3}\left(ea_{max}\right)^n\left(2^{2/3}+2^{1/6}\right)\right]^{\frac{1}{n+1}}, \quad (3.293)$$

$$t_{fa\,opt} = \left(18|e_0|\right)^{1/3}\left[\frac{|e_0|^{(5-n)/3}}{qn}\left(3\sqrt{2}\right)^{n/3} 3^{1/3}\left(ea_{max}\right)^n\left(2^{2/3}+2^{1/6}\right)\right]^{\frac{-1}{2n+2}}. \quad (3.294)$$

Criterion $Q^a_{ITAE}(k,|B|)$ as a function of two variables is presented in figure 3.69.

Fig. 3.69 The modified criterion $Q^a_{ITAE}(k,|B|)$

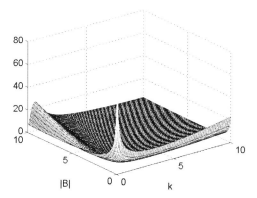

3.3.3 Switching Plane Design Subject to Velocity Constraint

Now we consider system (3.1) subject to velocity constraint (2.89). Taking into account the analysis presented at the beginning of section 3.2.3 we conclude that if system (3.1) is controlled according to the strategy proposed in this monograph, then constraint (2.89) is equivalent to relation (3.151). Thus, substituting (3.151) into criterion (3.216), we obtain the following single variable function which will be minimised further in the section

$$J^v_{ITAE}(k) = \frac{|e_0|^3}{v^2_{max}}\exp\left(\frac{-2ke^k}{e^k-1}\right)\left(\frac{1}{k}+\frac{1}{6}+\frac{3}{k^2}\right)\left(e^k-1\right)^2. \quad (3.295)$$

Criterion (3.295) is shown in figure 3.70. In a similar way, as we did in section 3.2.3, we can prove the following theorem.

Theorem 3
There exists such a value $k_{v\,max} \in (2, 2.2)$ that for any $k \in [0, k_{v\,max})$ function $J^v_{ITAE}(k)$, expressed by (3.295), increases, and it decreases for every $k \in (k_{v\,max}, \infty)$.

3.3 Switching Plane Design Minimising ITAE

Fig. 3.70 Criterion $J^v_{ITAE}(k)$

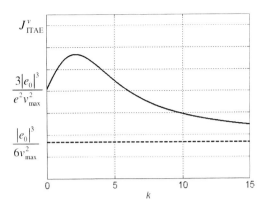

Proof

The derivative of function $J^v_{ITAE}(k)$ has the following form

$$\frac{dJ^v_{ITAE}(k)}{dk} = \frac{|e_0|^3}{v^2_{max} k^3} \exp\left(\frac{-2ke^k}{e^k-1}\right)\left[e^k\left(\frac{1}{3}k^4 + 2k^3 + 6k^2 + 2k + 12\right) - e^{2k}(k+6) - k - 6\right]. \quad (3.296)$$

Using Euler's sequence, we obtain

$$\frac{dJ^v_{ITAE}(k)}{dk} = \frac{|e_0|^3 e^k}{v^2_{max} k^3} \exp\left(\frac{-2ke^k}{e^k-1}\right)\left[\frac{1}{3}k^4 + 2k^3 + 6k^2 + 2k + 12 + \right.$$

$$-e^k(k+6) - e^{-k}(k+6)\right] = \frac{|e_0|^3 e^k}{v^2_{max} k^3} \exp\left(\frac{-2ke^k}{e^k-1}\right)\left[\frac{1}{3}k^4 + 2k^3 + 6k^2 + 2k + \right.$$

$$+12 - \left(1 + k + \frac{k^2}{2} + \frac{k^3}{6} + \frac{k^4}{24} + \frac{k^5}{120} + \ldots\right)(k+6) +$$

$$\left. -\left(1 - k + \frac{k^2}{2} - \frac{k^3}{6} + \frac{k^4}{24} - \frac{k^5}{120} + \ldots\right)(k+6)\right] = \quad (3.297)$$

$$= \frac{|e_0|^3 e^k}{v^2_{max} k^3} \exp\left(\frac{-2ke^k}{e^k-1}\right)\left(\frac{1}{3}k^4 + 2k^3 + 6k^2 + 2k + 12 + \right.$$

$$\left. -2k - k^3 - \frac{k^5}{12} - 12 - 6k^2 - \frac{k^4}{2} - \ldots\right) =$$

$$= \frac{|e_0|^3 e^k}{v^2_{max}} \exp\left(\frac{-2ke^k}{e^k-1}\right)\left(1 - \frac{k}{6} - \frac{k^2}{12} - \frac{k^3}{60} - \frac{k^4}{360} - \ldots\right).$$

It can be seen from (3.297) that $dJ_{ITAE}^{v}(k)/dk$ has only one root and that for any k smaller than the root, $dJ_{ITAE}^{v}(k)/dk$ is positive, and for any k greater than the root the derivative is negative. Let us denote the root as $k_{v\,max}$. Then

$$\left.\frac{dJ_{ITAE}^{v}(k)}{dk}\right|_{k=k_{v\,max}} = 0 \tag{3.298}$$

and from equations (3.296), (3.297) we obtain

$$\left.\frac{dJ_{ITAE}^{v}(k)}{dk}\right|_{k=2} > 0 \tag{3.299}$$

and

$$\left.\frac{dJ_{ITAE}^{v}(k)}{dk}\right|_{k=2.2} < 0 \tag{3.300}$$

which shows that as distinct from Theorem 1, we obtain $2 < k_{v\,max} < 2.2$. Hence, there exists such a value $k_{v\,max} \in (2, 2.2)$, for which function $J_{ITAE}^{v}(k)$ reaches its maximum. This ends the proof.

Taking into account Theorem 3 and the following relation

$$\lim_{k\to\infty} J_{ITAE}^{v}(k) = \frac{|e_0|^3}{6v_{max}^2} < \frac{3|e_0|^3}{e^2 v_{max}^2} = \lim_{k\to 0^+} J_{ITAE}^{v}(k), \tag{3.301}$$

we conclude that criterion (3.295) achieves its minimum value when $k_{v\,opt} \to \infty$. Then $B_{v\,opt} \to \mathrm{sgn}(e_0)\cdot\infty$, $A_{v\,opt} \to -\mathrm{sgn}(e_0)\cdot\infty$, $c_{1\,v\,opt} \to \infty$, $c_{2\,v\,opt} \to \infty$ and $t_{f\,v\,opt}$ is given by relation (2.98).

This solution is exactly the same as the one obtained in section 3.2.3. As expected, no matter which criterion – IAE or ITAE – is minimised, when only velocity constraint is considered, the system is supposed to move with its maximum admissible velocity right from the beginning of the control process. Furthermore, the system is also required to stop instantaneously, i.e. no slow down period is allowed. As we have already pointed out in section 3.2.3 such a motion cannot be realized (it would require infinite control power), and therefore, in the next

3.3 Switching Plane Design Minimising ITAE

sections we will consider the same system but subject to more than one constraint at the same time.

However, now we modify (in a similar way as in section 3.2.3) the considered criterion and by this means we introduce the elastic velocity constraint. This modified criterion has the following form

$$Q_{\text{ITAE}}^{v} = \int_{t_0}^{\infty} t \left| e_1(t) \right| dt + q \left\{ \frac{\max \left[\left| e_2(t) \right| \right]}{v_{\max}} \right\}^n \tag{3.302}$$

where as before $q > 0$ is a weighting factor and $n \geq 1$ is a constant determining how elastic the constraint is. Furthermore, $v_{\max}$ in equation (3.302) represents the threshold value of the system velocity. Substituting ITAE criterion given by (3.216) into relation (3.302) we get

$$Q_{\text{ITAE}}^{v} = \frac{\left| e_0 \right|^{5/3}}{\left| B \right|^{2/3}} \left(k^{1/3} + \frac{1}{6} k^{4/3} + 3 k^{-2/3} \right) + q \left\{ \frac{\max \left[\left| e_2(t) \right| \right]}{v_{\max}} \right\}^n . \tag{3.303}$$

Now we minimise criterion (3.303). Considering that $\max[|e_2(t)|]$ is described by the absolute value of (3.149), criterion Q_{ITAE}^{v} can be expressed as

$$Q_{\text{ITAE}}^{v}\left(k, \left| B \right|\right) = \frac{\left| e_0 \right|^{5/3}}{\left| B \right|^{2/3}} \left(k^{1/3} + \frac{1}{6} k^{4/3} + 3 k^{-2/3} \right) + \tag{3.304}$$

$$+ q \left[\exp \left(\frac{-k e^k}{e^k - 1} \right) \left(e^k - 1 \right) \frac{\left| B \right|^{1/3} e_0^{2/3}}{k^{2/3} v_{\max}} \right]^n .$$

Notice that for any k, derivative

$$\frac{\partial Q_{\text{ITAE}}^{v}\left(k, \left| B \right|\right)}{\partial \left| B \right|} = -\frac{2 \left| e_0 \right|^{5/3}}{3 \left| B \right|^{5/3}} \left(k^{1/3} + \frac{1}{6} k^{4/3} + 3 k^{-2/3} \right) + \tag{3.305}$$

$$+ \frac{1}{3} q n \left| B \right|^{n/3 - 1} \left[\exp \left(\frac{-k e^k}{e^k - 1} \right) \left(e^k - 1 \right) \frac{e_0^{2/3}}{k^{2/3} v_{\max}} \right]^n$$

is negative for small values of $|B|$, and for greater values of this parameter the derivative is getting positive. As the result of solving equation $\partial Q_{\text{ITAE}}^{v}\left(k, \left| B \right|\right) / \partial \left| B \right| = 0$ we obtain

$$|B| = \left\{ \frac{2|e_0|^{5/3}\left(k^{1/3} + \dfrac{1}{6}k^{4/3} + 3k^{-2/3}\right)}{qn\left[\exp\left(\dfrac{-ke^k}{e^k-1}\right)(e^k-1)\dfrac{e_0^{2/3}}{k^{2/3}v_{max}}\right]^n} \right\}^{\frac{3}{n+2}}. \tag{3.306}$$

Thus, for any k, criterion $Q_{ITAE}^v\left(k,|B|\right)$ reaches its minimum for $|B|$ presented by (3.306). Then, substituting (3.306) into criterion (3.304) we obtain

$$Q_{ITAE}^v\left(k,|B|\right) = \left(|e_0|^{5/3}\right)^{\frac{n}{n+2}}\left(\frac{qn}{2}\right)^{\frac{2}{n+2}}\left[\exp\left(\frac{-ke^k}{e^k-1}\right)(e^k-1)\frac{e_0^{2/3}}{k^{2/3}v_{max}}\right]^{\frac{2n}{n+2}}$$

$$\cdot\left(k^{1/3} + \frac{1}{6}k^{4/3} + 3k^{-2/3}\right)^{\frac{n}{n+2}} +$$

$$+q\left[\exp\left(\frac{-ke^k}{e^k-1}\right)(e^k-1)\frac{e_0^{2/3}}{k^{2/3}v_{max}}\right]^{\frac{2n}{n+2}}\left[\frac{2}{qn}|e_0|^{5/3}\left(k^{1/3} + \frac{1}{6}k^{4/3} + 3k^{-2/3}\right)\right]^{\frac{n}{n+2}} =$$

$$=\left(\frac{qn}{2}\right)^{\frac{2}{n+2}}\left[|e_0|^{5/3}\left(k^{1/3} + \frac{1}{6}k^{4/3} + 3k^{-2/3}\right)\exp\left(\frac{-2ke^k}{e^k-1}\right)(e^k-1)^2\frac{e_0^{4/3}}{k^{4/3}v_{max}^2}\right]^{\frac{n}{n+2}}.$$

$$\cdot\left(1+\frac{2}{n}\right) = \left(\frac{qn}{2}\right)^{\frac{2}{n+2}}\left[\frac{|e_0|^3}{v_{max}^2}\left(\frac{1}{k} + \frac{1}{6} + \frac{3}{k^2}\right)\exp\left(\frac{-2ke^k}{e^k-1}\right)(e^k-1)^2\right]^{\frac{n}{n+2}}\left(1+\frac{2}{n}\right). \tag{3.307}$$

From the above considerations and the analysis of the minimum of criterion (3.295), we conclude that function (3.307) achieves its minimum when k tends to infinity. Then substituting $k_{v\,opt} \to \infty$ into relation (3.306) we can see that the optimal value of parameter $|B|$ also tends to infinity. Thus, $B_{v\,opt} \to \mathrm{sgn}(e_0)\cdot\infty$, $A_{v\,opt} \to -\mathrm{sgn}(e_0)\cdot\infty$, $c_{1\,v\,opt} \to \infty$, $c_{2\,v\,opt} \to \infty$ and $t_{f\,v\,opt}$ is given by

$$t_{f\,v\,opt} = \left(\frac{3qn|e_0|^{n-1}}{v_{max}^n}\right)^{\frac{1}{n+2}}. \tag{3.308}$$

Comparing this equation and relation (3.166) one can easily notice that the optimal value $t_{fv\,opt}$ of the time when the switching plane stops moving determined in this section and in section 3.2.3 are expressed by two different formulae. This is partly due to the fact that the dimension of parameter q is not the same here and in

3.3 Switching Plane Design Minimising ITAE

section 3.2.3. Here the dimension is $(time)^2 \times error$ while in section 3.2.3 the dimension is $time \times error$.

That summarizes the analysis of the elastic velocity constraint. The plot of criterion $Q_{ITAE}^{v}(k,|B|)$ as a function of two variables is presented in figure 3.71. In this way we solved the task of the ITAE optimal switching plane parameters selection, when each of the constraints is taken into account separately. Further in this book, we keep minimising the same criterion but we move on to the case when more than one constraint has to be satisfied at the same time.

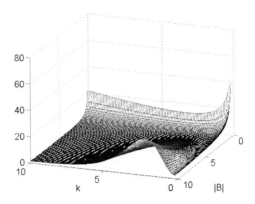

Fig. 3.71 The modified criterion $Q_{ITAE}^{v}(k,|B|)$

3.3.4 Switching Plane Design Subject to Acceleration and Velocity Constraints

In this section we take into account constraints (3.99) and (2.89) and we minimise ITAE with these two constraints taken into account simultaneously. Therefore, similarly as in section 3.2.4, for any k, the minimum value of $J_{ITAE}^{av}(k,B)$ is given by

$$J_{ITAE}^{av}(k) = \max\left[J_{ITAE}^{a}(k), J_{ITAE}^{v}(k)\right] \qquad (3.309)$$

where criteria $J_{ITAE}^{a}(k)$ and $J_{ITAE}^{v}(k)$ are given by functions (3.258) and (3.295), respectively. Consequently, this time the optimal solution of the minimisation of criterion $J_{ITAE}^{av}(k,B)$ is such a value of argument k, for which

$$J_{ITAE}^{av}(k_{av\,opt}) = \min_{k>0}\left\{J_{ITAE}^{av}(k)\right\} = \min_{k>0}\left\{\max\left[J_{ITAE}^{a}(k), J_{ITAE}^{v}(k)\right]\right\} \qquad (3.310)$$

and the respective value of B. Next, we will show the method for finding the optimal parameter $k_{av\,opt}$. In this method we need to consider two cases: $J_{ITAE}^{v}(3\sqrt{2}) \leq J_{ITAE}^{a}(3\sqrt{2})$ and $J_{ITAE}^{v}(3\sqrt{2}) > J_{ITAE}^{a}(3\sqrt{2})$. In the first case,

$k_{av\,opt} = k_{a\,opt} = 3\sqrt{2}$ and the optimal value of parameter B can be calculated from (3.260). Now we will consider the second case, illustrated in figure 3.72, i.e. the situation when $J^v_{ITAE}\left(3\sqrt{2}\right) > J^a_{ITAE}\left(3\sqrt{2}\right)$. Notice that the following theorem can be proved in the same way as Theorem 2, introduced in section 3.2.4.

Fig. 3.72 Criteria $J^v_{ITAE}(k)$ and $J^a_{ITAE}(k)$

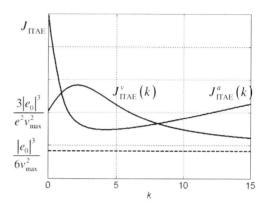

Theorem 4

If condition $J^v_{ITAE}\left(3\sqrt{2}\right) > J^a_{ITAE}\left(3\sqrt{2}\right)$ is satisfied, then criterion given by expression (3.309) achieves its minimum value at a point $k_{av\,opt}$ which belongs to the open interval from $3\sqrt{2}$ to $a_{max}e|e_0|/v^2_{max}$.

Proof

This proof corresponds to the proof of Theorem 2 with the difference that here we consider $J^a_{ITAE}(k)$, $J^v_{ITAE}(k)$ instead of $J^a_{IAE}(k)$, $J^v_{IAE}(k)$ and the critical point $3\sqrt{2}$ instead of 4. For any $k > 3\sqrt{2}$ criterion $J^v_{ITAE}(k)$ is a decreasing function of its argument and $J^a_{ITAE}(k)$ increases for any $k > 3\sqrt{2}$. Furthermore, the limit $\lim_{k \to \infty} J^a_{ITAE}(k) = \infty$. This means that there exists such a number k_α (because of similarity to the proof of Theorem 2 we take the same notation) which belongs to the interval $(3\sqrt{2}, \infty)$ that $J^a_{ITAE}(k_\alpha) = J^v_{ITAE}(k_\alpha)$. It can be demonstrated that function (3.309) achieves its minimum value at the point $k = k_\alpha$. For that purpose we consider two situations, i.e. we take into account $k_\alpha \in (3\sqrt{2}, 6)$ and $k_\alpha \geq 6$, which are similar to cases i) and iii) in the proof of Theorem 2.

i) case one $k_\alpha \in (3\sqrt{2}, 6)$

It follows from equation (3.258) that if $k_\alpha \leq 18$, then there exists such a number $k_\beta = 18/k_\alpha \geq 1$, that $J^a_{ITAE}(k_\alpha) = J^a_{ITAE}(k_\beta)$. Since $k_\alpha \in (3\sqrt{2}, 6)$, then

3.3 Switching Plane Design Minimising ITAE

$k_\beta \in (3, 3\sqrt{2})$. Then for any $k \notin (k_\beta, k_\alpha)$ we have $J^a_{\text{ITAE}}(k) > J^a_{\text{ITAE}}(k_\alpha) = J^a_{\text{ITAE}}(k_\beta)$. For any $k \in (k_{v\max}, k_\alpha)$, $J^v_{\text{ITAE}}(k) > J^v_{\text{ITAE}}(k_\alpha) = = J^a_{\text{ITAE}}(k_\alpha)$ which means that if $k_\beta > k_{v\max}$, then function $J^{av}_{\text{ITAE}}(k)$ has its minimum at the point k_α. Notice that $k_\beta > 3 > 2.2 > k_{v\max}$ and therefore function $J^{av}_{\text{ITAE}}(k)$ indeed achieves its minimum value at the point k_α.

ii) case two $k_\alpha \geq 6$

First we notice (as in the proof of Theorem 2) that function $J^v_{\text{ITAE}}(k)$ is continuous and decreasing for any $k > k_{v\max}$. On the other hand we have that $k_{v\max} < 3\sqrt{2}$, furthermore from (3.295)

$$J^v_{\text{ITAE}}\left(3\sqrt{2}\right) = \frac{|e_0|^3}{v^2_{\max}} \exp\left(\frac{-6\sqrt{2}e^{3\sqrt{2}}}{e^{3\sqrt{2}}-1}\right)\left(\frac{1}{3\sqrt{2}}+\frac{1}{3}\right)\left(e^{3\sqrt{2}}-1\right)^2 > \frac{3}{e^2}\frac{|e_0|^3}{v^2_{\max}} = \tag{3.311}$$
$$= \lim_{k \to 0^+} J^v_{\text{ITAE}}(k)$$

and

$$\lim_{k \to \infty} J^v_{\text{ITAE}}(k) = \frac{|e_0|^3}{6v^2_{\max}}, \tag{3.312}$$

we conclude that there exists exactly one number $z \in (3\sqrt{2}, \infty)$, such that for $k \in (3\sqrt{2}, \infty)$ we have

$$J^v_{\text{ITAE}}(k) < \frac{3}{e^2}\frac{|e_0|^3}{v^2_{\max}} \Leftrightarrow k > z. \tag{3.313}$$

Therefore, from Theorem 3 and the following relation

$$\lim_{k \to 0^+} J^v_{\text{ITAE}}(k) = \frac{3}{e^2}\frac{|e_0|^3}{v^2_{\max}}, \tag{3.314}$$

it follows that $J^v_{\text{ITAE}}(k)$ increases for any $k \in [0, k_{v\max})$ and consequently if $k > z$, then for any $k \in [0, k_{v\max})$, $J^v_{\text{ITAE}}(k) > J^v_{\text{ITAE}}(k_\alpha)$. On the other hand, $J^v_{\text{ITAE}}(k)$ decreases for any $k \in (k_{v\max}, \infty)$, so for any $k \in [k_{v\max}, k_\alpha)$, $J^v_{\text{ITAE}}(k) > J^v_{\text{ITAE}}(k_\alpha)$. This implies, that if $k_\alpha > z$, then $J^v_{\text{ITAE}}(k)$ achieves its minimum value at the point k_α. It is easy to verify that $z < 6$. Actually, the following inequality holds

$$J_{\text{ITAE}}^{v}(6) \approx 0.402 \frac{|e_0|^3}{v_{\text{max}}^2} < \frac{3}{e^2} \frac{|e_0|^3}{v_{\text{max}}^2}. \qquad (3.315)$$

Therefore, if $k_\alpha \geq 6$, then function $J_{\text{ITAE}}^{v}(k)$ has its minimum value at the point k_α.

The second part of this proof is concerned with demonstrating that $k_\alpha < a_{\text{max}} e|e_0|/v_{\text{max}}^2$. It can be verified (compare the proof of Theorem 2), that for any $k > 0$ the following inequality

$$\tilde{J}_{\text{ITAE}}^{v}(k) = \frac{|e_0|^3}{v_{\text{max}}^2} \left(\frac{1}{k} + \frac{1}{6} + \frac{3}{k^2} \right) >$$

$$> \frac{|e_0|^3}{v_{\text{max}}^2} \exp\left(\frac{-2ke^k}{e^k - 1} \right) \left(\frac{1}{k} + \frac{1}{6} + \frac{3}{k^2} \right) \left(e^k - 1 \right)^2 = J_{\text{ITAE}}^{v}(k) \qquad (3.316)$$

is satisfied. Therefore, taking into account that $J_{\text{ITAE}}^{v}\left(3\sqrt{2}\right) > J_{\text{ITAE}}^{a}\left(3\sqrt{2}\right)$, $\lim_{k \to \infty} J_{\text{ITAE}}^{a}(k) = \infty$ and $\lim_{k \to \infty} \tilde{J}_{\text{ITAE}}^{v}(k) = \frac{1}{6} \frac{|e_0|^3}{v_{\text{max}}^2}$, we conclude that there exists such a number $k_\gamma > 3\sqrt{2}$, that $\tilde{J}_{\text{ITAE}}^{v}(k_\gamma) = J_{\text{ITAE}}^{a}(k_\gamma)$. Solving equation

$$\frac{|e_0|^3}{v_{\text{max}}^2} \left(\frac{1}{k} + \frac{1}{6} + \frac{3}{k^2} \right) = \frac{e_0^2}{a_{\text{max}} e} \left(1 + \frac{1}{6} k + 3k^{-1} \right), \qquad (3.317)$$

we obtain its root $k_\gamma = a_{\text{max}} e|e_0|/v_{\text{max}}^2$. Then, since $\tilde{J}_{\text{ITAE}}^{v}$ dominates J_{ITAE}^{v}, we conclude that the optimal solution of criterion (3.309) minimisation task $k_{av\ opt} \in (3\sqrt{2}, k_\gamma)$. This ends the proof of Theorem 4.

From Theorem 4 we know that the optimal value $k_{av\ opt}$ of the parameter k belongs to the interval $(3\sqrt{2}, a_{\text{max}} e|e_0|/v_{\text{max}}^2)$ and that at this point $J_{\text{ITAE}}^{a}(k_{av\ opt}) = J_{\text{ITAE}}^{v}(k_{av\ opt})$. Moreover, for any $k \in (3\sqrt{2}, a_{\text{max}} e|e_0|/v_{\text{max}}^2)$ criterion $J_{\text{ITAE}}^{a}(k)$ is an increasing function of k and $J_{\text{ITAE}}^{v}(k)$ is a decreasing function of its argument. Therefore, in order to find the optimal value $k_{av\ opt}$ we introduce the following function

$$f_7(k) = J_{\text{ITAE}}^{v}(k) - J_{\text{ITAE}}^{a}(k) =$$

$$= \frac{e_0^2}{v_{\text{max}}^2 a_{\text{max}} e} \left(\frac{1}{k} + \frac{1}{6} + \frac{3}{k^2} \right) \left[a_{\text{max}} e|e_0| \exp\left(\frac{-2ke^k}{e^k - 1} \right) \left(e^k - 1 \right)^2 - v_{\text{max}}^2 k \right] \qquad (3.318)$$

which is monotonic in the considered interval. Notice that $f_7(3\sqrt{2}) \cdot f_7(a_{\text{max}} e|e_0|/v_{\text{max}}^2) < 0$. Therefore, $k_{av\ opt}$ which is the only root of equation

3.3 Switching Plane Design Minimising ITAE

$f_7(k) = 0$ in the interval $(3\sqrt{2}, a_{max}e|e_0|/v_{max}^2)$ can be easily found using any standard numerical procedure. The respective optimal value of B can be calculated from equations (3.112) or (3.159). The parameters determined in this way ensure the optimal performance of the controlled system together with satisfaction of both acceleration and velocity constraints.

It is worth to underline that using the presented method we obtain the same switching plane parameters as in section 3.2.4. Notice that the root of equation $f_7(k) = 0$ is exactly the same as the one obtained solving equation $f_3(k) = 0$ where $f_3(k)$ is given by (3.180). This can be seen from the following relation

$$
\begin{aligned}
f_7(k) &= \frac{e_0^2}{v_{max}^2 a_{max} e}\left(\frac{1}{k}+\frac{1}{6}+\frac{3}{k^2}\right)\left[a_{max} e|e_0|\exp\left(\frac{-2ke^k}{e^k-1}\right)\left(e^k-1\right)^2-v_{max}^2 k\right] = \\
&= \frac{e_0^2}{v_{max}^2 a_{max} e}\left(\frac{1}{k}+\frac{1}{6}+\frac{3}{k^2}\right)\left[\sqrt{a_{max} e|e_0|}\exp\left(\frac{-ke^k}{e^k-1}\right)\left(e^k-1\right)-v_{max}\sqrt{k}\right] \cdot \\
&\quad \cdot \left[\sqrt{a_{max} e|e_0|}\exp\left(\frac{-ke^k}{e^k-1}\right)\left(e^k-1\right)+v_{max}\sqrt{k}\right] = \\
&= \frac{\sqrt{|e_0|}}{v_{max}\sqrt{a_{max}} e}\frac{k^2+6k+18}{3k(k+4)}\left[\sqrt{a_{max} e|e_0|}\exp\left(\frac{-ke^k}{e^k-1}\right)\left(e^k-1\right)+v_{max}\sqrt{k}\right]f_3(k).
\end{aligned}
$$

$$(3.319)$$

The relation between the IAE and ITAE optimal solutions, when acceleration and velocity constraints are taken into account, is shown in table 3.1. Parameter $k_\gamma = a_{max}e|e_0|/v_{max}^2$ in this table is defined by (3.179) and its characteristic values are specified as follows

$$
k_{\gamma 1} = 4\exp\left(\frac{8e^4}{e^4-1}\right)\left(e^4-1\right)^{-2} \approx 4.819,
$$

$$(3.320)$$

Table 3.1 Relation between IAE and ITAE optimal solutions when acceleration and velocity constraints are considered

k_γ	$k_\gamma \in (0, k_{\gamma 1}]$	$k_\gamma \in (k_{\gamma 1}, k_{\gamma 2}]$	$k_\gamma > k_{\gamma 2}$
IAE optimal $k_{av\ opt}$	4	determined numerically $k_{av\ opt} \in (4, 3\sqrt{2}]$	determined numerically $k_{av\ opt} \in (3\sqrt{2}, k_\gamma)$
ITAE optimal $k_{av\ opt}$	$3\sqrt{2}$	$3\sqrt{2}$	$k_{av\ opt} \in (3\sqrt{2}, k_\gamma)$
J_{IAE}^v vs. J_{IAE}^a	$J_{IAE}^v(4) \leq J_{IAE}^a(4)$	$J_{IAE}^v(4) > J_{IAE}^a(4)$	$J_{IAE}^v(4) > J_{IAE}^a(4)$
J_{ITAE}^v vs. J_{ITAE}^a	$J_{ITAE}^v(3\sqrt{2}) < $ $< J_{ITAE}^a(3\sqrt{2})$	$J_{ITAE}^v(3\sqrt{2}) \leq$ $\leq J_{ITAE}^a(3\sqrt{2})$	$J_{ITAE}^v(3\sqrt{2}) >$ $> J_{ITAE}^a(3\sqrt{2})$

$$k_{\gamma 2} = 3\sqrt{2}\exp\left(\frac{6\sqrt{2}e^{3\sqrt{2}}}{e^{3\sqrt{2}}-1}\right)\left(e^{3\sqrt{2}}-1\right)^{-2} \approx 4.942. \tag{3.321}$$

3.3.5 Switching Plane Design Subject to Acceleration and Input Signal Constraints

In this section we will consider system (3.1) subject to the acceleration and input signal constraints. The greatest acceptable value of the input signal is denoted as $u_{\max}$ and the maximum admissible acceleration is $a_{\max}$. The following analysis is, to some extent, similar to the one presented in section 3.2.5. First, we notice that for any k, the minimum value of $J_{\mathrm{ITAE}}^{ua}(k,B)$ may be expressed as

$$J_{\mathrm{ITAE}}^{ua}(k) = \max\left[J_{\mathrm{ITAE}}^{u}(k), J_{\mathrm{ITAE}}^{a}(k)\right] \tag{3.322}$$

where criterion $J_{\mathrm{ITAE}}^{u}(k)$ is given by (3.217) and criterion $J_{\mathrm{ITAE}}^{a}(k)$ by (3.258). Consequently, function $J_{\mathrm{ITAE}}^{ua}(k,B)$ achieves its minimum for such a value of the argument k, for which

$$J_{\mathrm{ITAE}}^{ua}(k_{ua\,opt}) = \min_{k>0}\left\{\max\left[J_{\mathrm{ITAE}}^{u}(k), J_{\mathrm{ITAE}}^{a}(k)\right]\right\} \tag{3.323}$$

and a corresponding value of parameter B. Further in this section, the optimal parameter $k_{ua\,opt}$ will be determined in a similar way as in section 3.2.5. We consider the following three options:

- $J_{\mathrm{ITAE}}^{a}(k_{u\,opt}) \le J_{\mathrm{ITAE}}^{u}(k_{u\,opt})$,
- $J_{\mathrm{ITAE}}^{a}(k_{a\,opt}) \ge J_{\mathrm{ITAE}}^{u}(k_{a\,opt})$,
- $J_{\mathrm{ITAE}}^{a}(k_{u\,opt}) > J_{\mathrm{ITAE}}^{u}(k_{u\,opt}) \wedge J_{\mathrm{ITAE}}^{a}(k_{a\,opt}) < J_{\mathrm{ITAE}}^{u}(k_{a\,opt})$.

Case 1: $J_{\mathrm{ITAE}}^{a}(k_{u\,opt}) \le J_{\mathrm{ITAE}}^{u}(k_{u\,opt})$. In this case we have

$$\min_{k>0}\left\{\max\left[J_{\mathrm{ITAE}}^{u}(k), J_{\mathrm{ITAE}}^{a}(k)\right]\right\} = J_{\mathrm{ITAE}}^{u}(k_{u\,opt}). \tag{3.324}$$

This leads to the conclusion that the optimal pair $k_{ua\,opt} = k_{u\,opt}$ and $B_{ua\,opt} = B_{u\,opt}$ can be determined from (3.223) and (3.224).

Case 2: $J_{\mathrm{ITAE}}^{a}(k_{a\,opt}) \ge J_{\mathrm{ITAE}}^{u}(k_{a\,opt})$. In this situation

$$\min_{k>0}\left\{\max\left[J_{\mathrm{ITAE}}^{u}(k), J_{\mathrm{ITAE}}^{a}(k)\right]\right\} = J_{\mathrm{ITAE}}^{a}(k_{a\,opt}) = J_{\mathrm{ITAE}}^{a}\left(3\sqrt{2}\right). \tag{3.325}$$

3.3 Switching Plane Design Minimising ITAE

Consequently, we obtain $k_{ua\,opt} = k_{a\,opt} = 3\sqrt{2}$ and $B_{ua\,opt} = B_{a\,opt}$ given by equation (3.260).

Case 3: $J_{ITAE}^{a}\left(k_{u\,opt}\right) > J_{ITAE}^{u}\left(k_{u\,opt}\right) \wedge J_{ITAE}^{a}\left(k_{a\,opt}\right) < J_{ITAE}^{u}\left(k_{a\,opt}\right)$. In order to facilitate our further discussion, the optimisation task considered in this case is illustrated in figure 3.73.

Fig. 3.73 Criteria $J_{ITAE}^{u}(k)$ and $J_{ITAE}^{a}(k)$

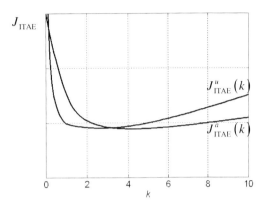

Let us first notice that the analysis presented in sections 3.3.1 and 3.3.2 shows that for any $k \in (k_{u\,opt}, k_{a\,opt})$, $J_{ITAE}^{u}(k)$ is an increasing function of k and $J_{ITAE}^{a}(k)$ is a decreasing function of its argument. Taking into account that $J_{ITAE}^{a}\left(k_{u\,opt}\right) > J_{ITAE}^{u}\left(k_{u\,opt}\right)$ and $J_{ITAE}^{a}\left(k_{a\,opt}\right) < J_{ITAE}^{u}\left(k_{a\,opt}\right)$, we conclude that the optimal value $k_{ua\,opt}$ of the parameter k belongs to the interval $(k_{u\,opt}, k_{a\,opt})$, where $k_{u\,opt} \approx 2.706$ and $k_{a\,opt} = 3\sqrt{2}$. At this point $J_{ITAE}^{u}\left(k_{ua\,opt}\right) = J_{ITAE}^{a}\left(k_{ua\,opt}\right)$. Consequently, in order to find the optimal value $k_{ua\,opt}$ we consider the following function

$$f_8(k) = J_{ITAE}^{u}(k) - J_{ITAE}^{a}(k) =$$
$$= |e_0|^{5/3}\left(k + \frac{1}{6}k^2 + 3\right)\left\{\left[\frac{e^{-k}(k-1)+1}{\delta U k}\right]^{2/3} - \frac{|e_0|^{1/3}}{a_{max}\,ek}\right\}. \tag{3.326}$$

This function is monotonic in the interval $(k_{u\,opt}, k_{a\,opt})$ and furthermore, $f_8(k_{a\,opt}) \cdot f_8(k_{u\,opt}) < 0$. Thus, the optimal parameter $k_{ua\,opt}$ can be numerically found from equation $f_8(k) = 0$. Next, the corresponding optimal value $B_{ua\,opt}$ may be calculated from any of relations (3.54) and (3.112). Notice that function $f_8(k)$ can be written as

$$f_8(k) = |e_0|^{5/3}\left(k + \frac{1}{6}k^2 + 3\right)\left\{\left[\frac{e^{-k}(k-1)+1}{\delta Uk}\right]^{2/3} - \frac{|e_0|^{1/3}}{a_{max}ek}\right\} =$$

$$= |e_0|^{5/3}\left(k + \frac{1}{6}k^2 + 3\right)\left\{\left[\frac{e^{-k}(k-1)+1}{\delta Uk}\right]^{1/3} - \frac{|e_0|^{1/6}}{\sqrt{a_{max}ek}}\right\}\cdot$$

$$\cdot\left\{\left[\frac{e^{-k}(k-1)+1}{\delta Uk}\right]^{1/3} + \frac{|e_0|^{1/6}}{\sqrt{a_{max}ek}}\right\} =$$

$$= \frac{|e_0|^{1/3}(k^2 + 6k + 18)}{3(4+k)}\left\{\left[\frac{e^{-k}(k-1)+1}{\delta Uk}\right]^{1/3} + \frac{|e_0|^{1/6}}{\sqrt{a_{max}ek}}\right\}f_4(k)$$

(3.327)

where $f_4(k)$ is given by equation (3.185). This means that the root of equation $f_8(k) = 0$ is the same as the root of equation $f_4(k) = 0$, which shows that for the fixed values of a_{max}, δ and U we will obtain the same value of optimal parameter $k_{ua\ opt}$ minimising criterion ITAE, as the one which minimises criterion IAE. This observation is in fact quite similar to our final remark concluding section 3.3.4.

The relation between the IAE and ITAE optimal solutions in the situation considered in this section, i.e. when input signal and acceleration constraints are analysed, is shown in table 3.2. By definition, parameter

$$\zeta = \frac{|e_0|(\delta U)^2}{(a_{max}e)^3}$$

(3.328)

and its distinctive values are given as follows

$$\zeta_1 = 2(e^{-2}+1)^2 \approx 2.577,$$

(3.329)

$$\zeta_2 = k_{u\ opt}^{ITAE}\left[e^{-k_{u\ opt}^{ITAE}}(k_{u\ opt}^{ITAE}-1)+1\right]^2 \approx 2.706(1.706e^{-2.706}+1)^2 \approx 3.358,$$

(3.330)

$$\zeta_3 = 4(3e^{-4}+1)^2 \approx 4.452,$$

(3.331)

$$\zeta_4 = 3\sqrt{2}\left[e^{-3\sqrt{2}}(3\sqrt{2}-1)+1\right]^2 \approx 4.647.$$

(3.332)

Symbol $k_{u\ opt}^{ITAE}$ in equation (3.330) denotes the value of k which is optimal in the sense of ITAE criterion with input signal constraint, i.e. $k_{u\ opt}^{ITAE} \approx 2.706$.

3.3 Switching Plane Design Minimising ITAE

Table 3.2 Relation between IAE and ITAE optimal solutions when acceleration and input signal constraints are considered

ζ	$\zeta \in (0,\zeta_1]$	$\zeta \in (\zeta_1,\zeta_2]$	$\zeta \in (\zeta_2,\zeta_3)$	$\zeta \in [\zeta_3,\zeta_4)$	$\zeta \geq \zeta_4$
IAE optimal $k_{ua\,opt}$	2	determined numerically $k_{ua\,opt} \in$ $(2, k_{u\,opt}^{ITAE}\,]$	determined numerically $k_{ua\,opt} \in$ $(k_{u\,opt}^{ITAE}\,, 4)$	4	4
ITAE optimal $k_{ua\,opt}$	$k_{u\,opt}^{ITAE}$	$k_{u\,opt}^{ITAE}$		determined numerically $k_{ua\,opt}$ $\in [4,\ 3\sqrt{2})$	$3\sqrt{2}$
J_{IAE}^{a} vs. J_{IAE}^{u}	$J_{IAE}^{a}(2) \leq$ $\leq J_{IAE}^{u}(2)$	$J_{IAE}^{a}(2) >$ $> J_{IAE}^{u}(2) \ \wedge$ $J_{IAE}^{a}(4) <$ $< J_{IAE}^{u}(4)$	$J_{IAE}^{a}(2) >$ $> J_{IAE}^{u}(2) \ \wedge$ $J_{IAE}^{a}(4) <$ $< J_{IAE}^{u}(4)$	$J_{IAE}^{a}(4) \geq$ $\geq J_{IAE}^{u}(4)$	$J_{IAE}^{a}(4) >$ $> J_{IAE}^{u}(4)$
J_{ITAE}^{a} vs. J_{ITAE}^{u}	$J_{ITAE}^{a}\left(k_{u\,opt}^{ITAE}\right) <$ $< J_{ITAE}^{u}\left(k_{u\,opt}^{ITAE}\right)$	$J_{ITAE}^{a}\left(k_{u\,opt}^{ITAE}\right) \leq$ $\leq J_{ITAE}^{u}\left(k_{u\,opt}^{ITAE}\right)$	$J_{ITAE}^{a}\left(k_{u\,opt}^{ITAE}\right) >$ $> J_{ITAE}^{u}\left(k_{u\,opt}^{ITAE}\right)$ $\wedge$ $J_{ITAE}^{a}\left(3\sqrt{2}\right) <$ $< J_{ITAE}^{u}\left(3\sqrt{2}\right)$	$J_{ITAE}^{a}\left(k_{u\,opt}^{ITAE}\right) >$ $> J_{ITAE}^{u}\left(k_{u\,opt}^{ITAE}\right)$ $\wedge$ $J_{ITAE}^{a}\left(3\sqrt{2}\right) <$ $< J_{ITAE}^{u}\left(3\sqrt{2}\right)$	$J_{ITAE}^{a}\left(3\sqrt{2}\right) \geq$ $\geq J_{ITAE}^{u}\left(3\sqrt{2}\right)$

3.3.6 *Switching Plane Design Subject to Input Signal and Velocity Constraints*

In this section we take into account the last pair of the three constraints considered in the monograph. We design the switching plane when u_{max} is the maximum admissible value of the input signal and the velocity cannot exceed v_{max}, i.e. we require inequalities (2.47) and (2.89) to be satisfied. In order to minimise criterion $J_{ITAE}(k,B)$ given by (3.216) we will find the minimum of the following function

$$J_{ITAE}^{uv}(k) = \max\left[J_{ITAE}^{u}(k), J_{ITAE}^{v}(k)\right]. \tag{3.333}$$

The optimal solution of the minimisation task will consist of such a value $k_{uv\,opt}$ of argument k which satisfies the following condition

$$J_{ITAE}^{uv}(k_{uv\,opt}) = \min_{k>0}\left\{\max\left[J_{ITAE}^{u}(k), J_{ITAE}^{v}(k)\right]\right\} \tag{3.334}$$

and the corresponding value of parameter B. We take into account the following two cases: $J^v_{\text{ITAE}}(k_{u\,opt}) \leq J^u_{\text{ITAE}}(k_{u\,opt})$ and $J^v_{\text{ITAE}}(k_{u\,opt}) > J^u_{\text{ITAE}}(k_{u\,opt})$. Notice that in the first case when $J^v_{\text{ITAE}}(k_{u\,opt}) \leq J^u_{\text{ITAE}}(k_{u\,opt})$ we obtain

$$\min_{k>0}\left\{\max\left[J^u_{\text{ITAE}}(k), J^v_{\text{ITAE}}(k)\right]\right\} = J^u_{\text{ITAE}}(k_{u\,opt}). \tag{3.335}$$

Therefore, the optimal value of k is $k_{uv\,opt} = k_{u\,opt} \approx 2.706$ and the optimal parameter $B_{uv\,opt} = B_{u\,opt}$ can be found from (3.224).

Fig. 3.74 Criteria $J^u_{\text{ITAE}}(k)$ and $J^v_{\text{ITAE}}(k)$

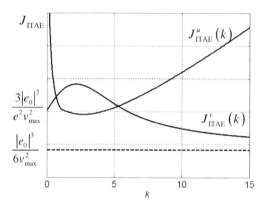

In the second case, illustrated in figure 3.74, i.e. when $J^v_{\text{ITAE}}(k_{u\,opt}) > J^u_{\text{ITAE}}(k_{u\,opt})$, criterion (3.334) achieves its minimum value at a point $k_{uv\,opt}$ which belongs to the open interval from $k_{u\,opt}$ to $p^{3/2}$ where p is given by (3.192). This statement can be justified as follows. Notice that for any $k > k_{u\,opt}$ criterion $J^v_{\text{ITAE}}(k)$ is a decreasing function of its argument and criterion $J^u_{\text{ITAE}}(k)$ is an increasing function of k. Furthermore, $\lim_{k\to\infty} J^u_{\text{ITAE}}(k) = \infty$ and in the case considered here $J^v_{\text{ITAE}}(k_{u\,opt}) > J^u_{\text{ITAE}}(k_{u\,opt})$. Hence, there exists such a point $k_\alpha \in (k_{u\,opt}, \infty)$ that $J^u_{\text{ITAE}}(k_\alpha) = J^v_{\text{ITAE}}(k_\alpha)$ and at this point criterion (3.333) has a local minimum. Furthermore, from the assumption $J^v_{\text{ITAE}}(k_{u\,opt}) > J^u_{\text{ITAE}}(k_{u\,opt})$, substituting $k = k_{u\,opt}$ into equations (3.217) and (3.295), we obtain

$$\frac{|e_0|^3}{v_{\max}^2}\exp\left(\frac{-2k_{u\,opt}e^{k_{u\,opt}}}{e^{k_{u\,opt}}-1}\right)\left(\frac{1}{k_{u\,opt}}+\frac{1}{6}+\frac{3}{k_{u\,opt}^2}\right)\left(e^{k_{u\,opt}}-1\right)^2 > \\ > \frac{|e_0|^{5/3}}{(\delta U)^{2/3}}\left[e^{-k_{u\,opt}}(k_{u\,opt}-1)+1\right]^{2/3}\left(k_{u\,opt}^{1/3}+\frac{1}{6}k_{u\,opt}^{4/3}+3k_{u\,opt}^{-2/3}\right) \tag{3.336}$$

3.3 Switching Plane Design Minimising ITAE

and

$$\frac{|e_0|^3}{v_{max}^2} > \frac{|e_0|^{5/3}}{(\delta U)^{2/3}} \frac{k_{u\,opt}^{4/3}}{\left(e^{k_{u\,opt}}-1\right)^2} \left[e^{-k_{u\,opt}}\left(k_{u\,opt}-1\right)+1\right]^{2/3} \exp\left(\frac{2k_{u\,opt}e^{k_{u\,opt}}}{e^{k_{u\,opt}}-1}\right). \quad (3.337)$$

Therefore,

$$J_{ITAE}^{v}(2.1) \approx 0.56703\frac{|e_0|^3}{v_{max}^2} > 3.87093\frac{|e_0|^{5/3}}{(\delta U)^{2/3}} \approx J_{ITAE}^{u}(2.1). \quad (3.338)$$

Moreover, for any $k < 2.1$ criterion $J_{ITAE}^{u}(k)$ decreases, $\lim_{k\to 0^+} J_{ITAE}^{u}(k) = \infty$ and for any $k < 2.1$ criterion $J_{ITAE}^{v}(k)$ is an increasing function of its argument. This means that there exists such a point $k \in (0, 2.1)$ that $J_{ITAE}^{u}(k_\chi) = J_{ITAE}^{v}(k_\chi)$ and at this point criterion (3.333) has another local minimum. It can be shown that function (3.333) achieves its global (for $k > 0$) minimum at the point $k = k_\alpha$, i.e. $J_{ITAE}^{u}(k_\chi) = J_{ITAE}^{v}(k_\chi) > J_{ITAE}^{v}(k_\alpha) = J_{ITAE}^{u}(k_\alpha)$. The proof of this property is similar to the proof of Theorem 2 however, it is much more tedious since it requires eight different cases to be considered. Because of this fact the details of the proof will not be presented here.

Now we will show how to find numerically the point $k_{uv\,opt} = k_\alpha$. Let us first notice that

$$\frac{|e_0|^{5/3}}{(\delta U)^{2/3}}\left(k^{1/3} + \frac{1}{6}k^{4/3} + 3k^{-2/3}\right) \le$$

$$\le \frac{|e_0|^{5/3}}{(\delta U)^{2/3}}\left[e^{-k}(k-1)+1\right]^{2/3}\left(k^{1/3} + \frac{1}{6}k^{4/3} + 3k^{-2/3}\right) = J_{ITAE}^{u}(k). \quad (3.339)$$

In this way we determined the lower bound of criterion $J_{ITAE}^{u}(k)$. On the other hand, we have

$$J_{ITAE}^{v}(k) = \frac{|e_0|^3}{v_{max}^2}\exp\left(\frac{-2ke^k}{e^k-1}\right)\left(\frac{1}{k}+\frac{1}{6}+\frac{3}{k^2}\right)(e^k-1)^2 < \frac{|e_0|^3}{v_{max}^2}\left(\frac{1}{k}+\frac{1}{6}+\frac{3}{k^2}\right) \quad (3.340)$$

This inequality specifies the upper bound of criterion $J_{ITAE}^{v}(k)$. Then solving equation

$$\frac{|e_0|^{5/3}}{(\delta U)^{2/3}}k^{4/3}\left(\frac{1}{k}+\frac{1}{6}+\frac{3}{k^2}\right) = \frac{|e_0|^3}{v_{max}^2}\left(\frac{1}{k}+\frac{1}{6}+\frac{3}{k^2}\right), \quad (3.341)$$

we obtain

$$k_\gamma = \frac{|e_0|\sqrt{\delta U}}{v_{max}^{3/2}} = \left[\frac{|e_0|^{2/3}(\delta U)^{1/3}}{v_{max}}\right]^{3/2} = p^{3/2},\tag{3.342}$$

and we conclude that $k_\alpha \in (k_{u\,opt}, p^{3/2})$. Then, in order to find $k_{uv\,opt}$, we calculate numerically the only root of the following function

$$f_9(k) = J_{ITAE}^v(k) - J_{ITAE}^u(k) = \frac{|e_0|^{5/3}}{(\delta U)^{2/3}}\left(\frac{1}{k}+\frac{1}{6}+\frac{3}{k^2}\right).$$

$$\cdot\left\{\frac{(\delta U)^{2/3}|e_0|^{4/3}}{v_{max}^2}\left(e^k-1\right)^2\exp\left(\frac{-2ke^k}{e^k-1}\right)-k^{4/3}\left[e^{-k}(k-1)+1\right]^{2/3}\right\}=$$

$$=\frac{|e_0|^{5/3}}{(\delta U)^{2/3}}\left(\frac{1}{k}+\frac{1}{6}+\frac{3}{k^2}\right).$$

$$\cdot\left\{\left[\frac{(\delta U)^{1/3}|e_0|^{2/3}}{v_{max}}\right]^2\left(e^k-1\right)^2\exp\left(\frac{-2ke^k}{e^k-1}\right)-k^{4/3}\left[e^{-k}(k-1)+1\right]^{2/3}\right\}=$$

$$=\frac{|e_0|^{5/3}}{(\delta U)^{2/3}}\left(\frac{1}{k}+\frac{1}{6}+\frac{3}{k^2}\right)\left\{p^2\left(e^k-1\right)^2\exp\left(\frac{-2ke^k}{e^k-1}\right)-k^{4/3}\left[e^{-k}(k-1)+1\right]^{2/3}\right\}$$

$$\tag{3.343}$$

in the interval $(k_{u\,opt}, p^{3/2})$. Finally, the optimal value $B_{uv\,opt}$ of parameter B can be found from equation (3.54) or (3.159).

The following relation

$$f_9(k) =$$

$$=\frac{|e_0|^{5/3}}{(\delta U)^{2/3}}\left(\frac{1}{k}+\frac{1}{6}+\frac{3}{k^2}\right)\left\{p^2\left(e^k-1\right)^2\exp\left(\frac{-2ke^k}{e^k-1}\right)-k^{4/3}\left[e^{-k}(k-1)+1\right]^{2/3}\right\}=$$

$$=\frac{|e_0|^{5/3}}{(\delta U)^{2/3}}\left(\frac{1}{k}+\frac{1}{6}+\frac{3}{k^2}\right)\left\{p\left(e^k-1\right)\exp\left(\frac{-ke^k}{e^k-1}\right)-k^{2/3}\left[e^{-k}(k-1)+1\right]^{1/3}\right\}.$$

$$\cdot\left\{p\left(e^k-1\right)\exp\left(\frac{-ke^k}{e^k-1}\right)+k^{2/3}\left[e^{-k}(k-1)+1\right]^{1/3}\right\}=$$

$$=\frac{|e_0|^{1/3}(k^2+6k+18)}{(\delta U)^{1/3}\,3k(4+k)}\left\{p\left(e^k-1\right)\exp\left(\frac{-ke^k}{e^k-1}\right)+k^{2/3}\left[e^{-k}(k-1)+1\right]^{1/3}\right\}f_6(k)$$

$$\tag{3.344}$$

3.3 Switching Plane Design Minimising ITAE

Table 3.3 Relation between IAE and ITAE optimal solutions when velocity and input signal constraints are considered

p	$p \in (0, p_0]$	$p \in (p_0, p_1]$	$p \in (p_1, p_2]$	$p > p_2$
IAE optimal $k_{uv\,opt}$	2	2	determined numerically $k_{uv\,opt} \in (0, 2)$	determined numerically $k_{uv\,opt} \in (k_{u\,opt}^{ITAE}, p^{3/2})$
ITAE optimal $k_{uv\,opt}$	$k_{u\,opt}^{ITAE}$	determined numerically $k_{uv\,opt} \in (k_{u\,opt}^{ITAE}, p^{3/2})$	determined numerically $k_{uv\,opt} \in (k_{u\,opt}^{ITAE}, p^{3/2})$	
J_{IAE}^{v} vs. J_{IAE}^{u}	$J_{IAE}^{v}(2) < < J_{IAE}^{u}(2)$	$J_{IAE}^{v}(2) \le \le J_{IAE}^{u}(2)$	$J_{IAE}^{v}(2) > > J_{IAE}^{u}(2)$	$J_{IAE}^{v}(2) > > J_{IAE}^{u}(2)$
J_{ITAE}^{v} vs. J_{ITAE}^{u}	$J_{IAE}^{v}(k_{u\,opt}^{ITAE}) \le \le J_{IAE}^{u}(k_{u\,opt}^{ITAE})$	$J_{IAE}^{v}(k_{u\,opt}^{ITAE}) > > J_{IAE}^{u}(k_{u\,opt}^{ITAE})$	$J_{IAE}^{v}(k_{u\,opt}^{ITAE}) > > J_{IAE}^{u}(k_{u\,opt}^{ITAE})$	$J_{IAE}^{v}(k_{u\,opt}^{ITAE}) > > J_{IAE}^{u}(k_{u\,opt}^{ITAE})$

where $f_6(k)$ is determined by equation (3.198), shows that the optimal parameter $k_{uv\,opt}$ obtained from the minimisation of criterion ITAE may be the same as $k_{uv\,opt}$ determined by the minimisation of criterion IAE.

The relation between the IAE and ITAE optimal solutions, with velocity and input signal constraints, is shown in table 3.3. In this table, similarly as in table 3.2, $k_{u\,opt}^{ITAE}$ represents the ITAE optimal value of k when input signal constraint is taken into account. Furthermore, in table 3.3 parameter $p = |e_0|^{2/3} (\delta U)^{1/3} / v_{max}$ is defined by (3.192). Moreover, $p_1 \approx 2.619$ given by (3.193) and $p_2 \approx 3.2685$ have been calculated in section 3.2.6. Another characteristic value of parameter p is presented below

$$p_0 = \frac{\left(k_{u\,opt}^{ITAE}\right)^{2/3} \left[e^{-k_{u\,opt}^{ITAE}} \left(k_{u\,opt}^{ITAE} - 1\right) + 1\right]^{1/3}}{e^{k_{u\,opt}^{ITAE}} - 1} \exp\left(\frac{k_{u\,opt}^{ITAE} e^{k_{u\,opt}^{ITAE}}}{e^{k_{u\,opt}^{ITAE}} - 1}\right) \approx$$

$$\approx \frac{(2.706)^{2/3} \left(1.706 \, e^{-2.706} + 1\right)^{1/3}}{e^{2.706} - 1} \exp\left(\frac{2.706 \, e^{2.706}}{e^{2.706} - 1}\right) \approx 2.618. \tag{3.345}$$

3.3.7 Switching Plane Design with Acceleration, Velocity and Input Signal Constraints

In this section, similarly as in section 3.2.7, we will consider system (3.1) with limited acceleration, input signal and velocity. The maximum admissible values of acceleration, the input signal and velocity are again denoted as a_{max}, u_{max} and v_{max}, respectively. In this section, in order to determine the optimal switching plane parameters, the minimum of the following criterion

$$J_{ITAE}^{uav}(k) = \max\left[J_{ITAE}^{u}(k), J_{ITAE}^{a}(k), J_{ITAE}^{v}(k)\right] \qquad (3.346)$$

will be found. Consequently, the optimal solution of the minimisation of criterion $J_{ITAE}(k, B)$ with constraints (2.47), (3.99) and (2.89) is such a pair (k_{opt}, B_{opt}) that

$$J_{ITAE}^{uav}(k_{opt}) = \min_{k>0}\left\{\max\left[J_{ITAE}^{u}(k), J_{ITAE}^{a}(k), J_{ITAE}^{v}(k)\right]\right\} \qquad (3.347)$$

and B_{opt} is a corresponding value of B. Further in this section, the algorithm for finding k_{opt} will be presented. We will consider the following options:

i) If $J_{ITAE}^{a}(k_{u\,opt}) \leq J_{ITAE}^{u}(k_{u\,opt}) \wedge J_{ITAE}^{v}(k_{u\,opt}) \leq J_{ITAE}^{u}(k_{u\,opt})$, then $k_{opt} = k_{u\,opt} \approx 2.706$ and the optimal value B_{opt} can be calculated from formula (3.224). This case is illustrated in figure 3.75.

ii) If $J_{ITAE}^{a}(k_{u\,opt}) \leq J_{ITAE}^{u}(k_{u\,opt}) \wedge J_{ITAE}^{v}(k_{u\,opt}) > J_{ITAE}^{u}(k_{u\,opt})$, then in order to find the optimal value of parameter k equation $J_{ITAE}^{v}(k) - J_{ITAE}^{u}(k) = 0$ (which is equivalent to equation $f_9(k) = 0$ where $f_9(k)$ is given by (3.343)) should be solved in the interval $(k_{u\,opt}, p^{3/2})$. Then parameter B_{opt} may be obtained from (3.54) as well as from (3.159). This way of finding the optimal solution is a direct consequence of the following implication

$$\underset{k_\theta \geq k_{u\,opt}}{\forall}\left[J_{ITAE}^{a}(k_\theta) \leq J_{ITAE}^{u}(k_\theta) \Rightarrow \underset{k \geq k_\theta}{\forall} J_{ITAE}^{a}(k) \leq J_{ITAE}^{u}(k)\right]. \qquad (3.348)$$

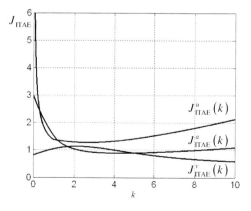

Fig. 3.75 Criteria $J_{ITAE}^{u}(k)$, $J_{ITAE}^{a}(k)$ and $J_{ITAE}^{v}(k)$ versus k

3.3 Switching Plane Design Minimising ITAE

In order to prove this implication we consider assumption $J^a_{ITAE}(k_\theta) \leq J^u_{ITAE}(k_\theta)$, which is equivalent to

$$\frac{e_0^2}{a_{max} e} \leq \frac{|e_0|^{5/3}}{(\delta U)^{2/3}} \left[e^{-k_\theta}(k_\theta - 1) + 1 \right]^{2/3} k_\theta^{1/3}. \qquad (3.349)$$

Next, we show that the following quotient $J^a_{ITAE}(k)/J^u_{ITAE}(k)$ is smaller than or equal to one for any $k \geq k_\theta \geq k_{u\,opt}$.

$$\frac{J^a_{ITAE}(k)}{J^u_{ITAE}(k)} = \frac{e_0^2}{a_{max} e} \frac{(\delta U)^{2/3}}{|e_0|^{5/3}} \left[e^{-k}(k-1)+1 \right]^{-2/3} k^{-1/3}. \qquad (3.350)$$

Then, substituting inequality (3.349) into equation (3.350) and considering the fact that expression $\left[e^{-k}(k-1)+1 \right]^{2/3} k^{1/3}$ is an increasing function of argument k, we get for any $k \geq k_\theta \geq k_{u\,opt}$

$$\frac{J^a_{ITAE}(k)}{J^u_{ITAE}(k)} \leq \frac{|e_0|^{5/3}}{(\delta U)^{2/3}} \left[e^{-k_\theta}(k_\theta - 1) + 1 \right]^{2/3} k_\theta^{1/3}$$
$$\cdot \frac{(\delta U)^{2/3}}{|e_0|^{5/3}} \left[e^{-k}(k-1)+1 \right]^{-2/3} k^{-1/3} \leq 1 \qquad (3.351)$$

which demonstrates that implication (3.348) is true. Condition (3.348) with $k_\theta = k_{u\,opt}$ justifies the proposed procedure of finding k_{opt} in the considered case. The analysis presented here is illustrated in figure 3.76.

Fig. 3.76 Criteria $J^u_{ITAE}(k)$, $J^a_{ITAE}(k)$ and $J^v_{ITAE}(k)$ versus k

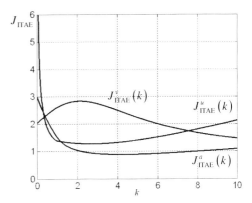

iii) If $J^a_{ITAE}\left(3\sqrt{2}\right) \geq J^u_{ITAE}\left(3\sqrt{2}\right) \wedge J^v_{ITAE}\left(3\sqrt{2}\right) \leq J^a_{ITAE}\left(3\sqrt{2}\right)$, then $k_{opt} = 3\sqrt{2}$ and parameter B_{opt} is presented by (3.260). This situation is shown in figure 3.77. Notice that the first assumption considered in this case, i.e. inequality $J^a_{ITAE}\left(3\sqrt{2}\right) \geq J^u_{ITAE}\left(3\sqrt{2}\right)$ implies that $J^a_{ITAE}\left(k_{u\,opt}\right) > J^u_{ITAE}\left(k_{u\,opt}\right)$.

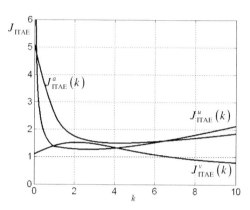

Fig. 3.77 Criteria $J^u_{ITAE}(k)$, $J^a_{ITAE}(k)$ and $J^v_{ITAE}(k)$ versus k

iv) If $J^a_{ITAE}\left(k_{u\,opt}\right) > J^u_{ITAE}\left(k_{u\,opt}\right) \wedge J^a_{ITAE}\left(3\sqrt{2}\right) \geq J^u_{ITAE}\left(3\sqrt{2}\right) \wedge$
$\wedge J^v_{ITAE}\left(3\sqrt{2}\right) > J^a_{ITAE}\left(3\sqrt{2}\right)$, then in order to find the optimal value of parameter k first we solve equation $J^v_{ITAE}(k) - J^u_{ITAE}(k) = 0$ (i.e. equation $f_9(k) = 0$) in the interval $(3\sqrt{2}, p^{3/2})$ and next equation $J^v_{ITAE}(k) - J^a_{ITAE}(k) = 0$ (which is equivalent to solving equation $f_7(k) = 0$ with $f_7(k)$ given by (3.318)) in the interval $(3\sqrt{2}, a_{max}e|e_0|/v^2_{max})$. As the optimal value of parameter k we choose this one of the two solutions which guarantees greater value of criterion $J^v_{ITAE}(k)$. Then, parameter B_{opt} may be found from equation (3.54) if k_{opt} is the root of the first equation, i.e. $J^v_{ITAE}(k) - J^u_{ITAE}(k) = 0$, and B_{opt} can be calculated from formula (3.112) when k_{opt} is determined solving equation $J^v_{ITAE}(k) - J^a_{ITAE}(k) = 0$. In both of these situations the optimal parameter B_{opt} is also given by (3.159). The optimisation task considered in this case is illustrated in figure 3.78.

v) If $J^a_{ITAE}\left(k_{u\,opt}\right) > J^u_{ITAE}\left(k_{u\,opt}\right) \wedge J^a_{ITAE}\left(3\sqrt{2}\right) < J^u_{ITAE}\left(3\sqrt{2}\right)$, then at the beginning we solve equation $J^u_{ITAE}(k) - J^a_{ITAE}(k) = 0$ in the interval

3.3 Switching Plane Design Minimising ITAE

Fig. 3.78 Criteria $J^u_{ITAE}(k)$, $J^a_{ITAE}(k)$ and $J^v_{ITAE}(k)$ versus k

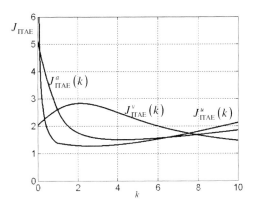

Fig. 3.79 Criteria $J^u_{ITAE}(k)$, $J^a_{ITAE}(k)$ and $J^v_{ITAE}(k)$ versus k

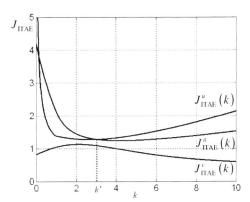

Fig. 3.80 Criteria $J^u_{ITAE}(k)$, $J^a_{ITAE}(k)$ and $J^v_{ITAE}(k)$ versus k

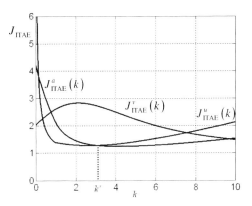

($k_{u\,opt}$, $3\sqrt{2}$) which is actually equivalent to equation $f_8(k) = 0$ where $f_8(k)$ is determined by (3.326). We denote the root of this equation as k^*. Next we consider the following two subcases:

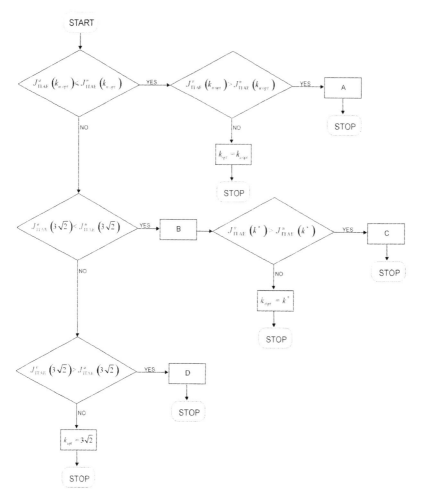

Fig. 3.81 The algorithm for finding k_{opt}

- If $J_{ITAE}^v\left(k^*\right) \leq J_{ITAE}^a\left(k^*\right) = J_{ITAE}^u\left(k^*\right)$, then $k_{opt} = k^*$. Next, the optimal value of parameter B can be calculated from equations (3.54) or (3.112). This scenario is depicted in figure 3.79.

- If $J_{ITAE}^v\left(k^*\right) > J_{ITAE}^a\left(k^*\right) = J_{ITAE}^u\left(k^*\right)$, then in order to obtain parameter k_{opt} we solve equation $J_{ITAE}^v\left(k\right) - J_{ITAE}^u\left(k\right) = 0$, i.e. equation $f_0(k) = 0$ in the interval $(k^*, p^{3/2})$. This is a direct consequence of implication (3.348) with $k = 3\sqrt{2}$. In this case parameter B_{opt} may be calculated from any of equations (3.54) or (3.159). This optimisation task is visualised in figure 3.80.

3.3 Switching Plane Design Minimising ITAE

The above method is illustrated by the block diagram, shown in figure 3.81. Procedures A, B, C, D are described below.

Procedures for finding parameter k_{opt} :

Procedure A: Solve equation

$$\frac{|e_0|^{5/3}}{(\delta U)^{2/3}}\left(\frac{1}{k}+\frac{1}{6}+\frac{3}{k^2}\right)\left\{p^2\left(e^k-1\right)^2\exp\left(\frac{-2\,ke^k}{e^k-1}\right)-k^{4/3}\left[e^{-k}\left(k-1\right)+1\right]^{2/3}\right\}=0$$

in the interval $(k_{u\,opt},\,p^{3/2})$.

Procedure B: Solve equation

$$|e_0|^{5/3}\left(k+\frac{1}{6}k^2+3\right)\left\{\left[\frac{e^{-k}\left(k-1\right)+1}{\delta Uk}\right]^{2/3}-\frac{|e_0|^{1/3}}{a_{max}\,ek}\right\}=0$$

in the interval $(k_{u\,opt},\,3\sqrt{2}\,)$. Denote the obtained root as k^*.

Procedure C: Solve equation

$$\frac{|e_0|^{5/3}}{(\delta U)^{2/3}}\left(\frac{1}{k}+\frac{1}{6}+\frac{3}{k^2}\right)\left\{p^2\left(e^k-1\right)^2\exp\left(\frac{2\,ke^k}{e^k-1}\right)-k^{4/3}\left[e^{-k}\left(k-1\right)+1\right]^{2/3}\right\}=0$$

in the interval $(k^*,\,p^{3/2})$.

Procedure D: Solve the following two equations. The first one

$$\frac{|e_0|^{5/3}}{(\delta U)^{2/3}}\left(\frac{1}{k}+\frac{1}{6}+\frac{3}{k^2}\right)\left\{p^2\left(e^k-1\right)^2\exp\left(\frac{-2\,ke^k}{e^k-1}\right)-k^{4/3}\left[e^{-k}\left(k-1\right)+1\right]^{2/3}\right\}=0$$

in the interval $(3\sqrt{2}\,,\,p^{3/2})$, and the second one

$$\frac{e_0^2}{v_{max}^2\,a_{max}\,e}\left(\frac{1}{k}+\frac{1}{6}+\frac{3}{k^2}\right)\left[a_{max}\,e|e_0|\exp\left(\frac{-2\,ke^k}{e^k-1}\right)\left(e^k-1\right)^2-v_{max}^2\,k\right]=0$$

in the interval $(3\sqrt{2}\,,\,a_{max}e|e_0|/v_{max}^2\,)$. Choose this solution which gives greater value of criterion $J_{ITAE}^v\left(k\right)$.

Chapter 4
Conclusions

Sliding mode control is well known to be a robust and computationally efficient regulation technique which may be applied to nonlinear and possibly time-varying plants. Therefore, the proper design of the sliding mode controllers has recently become one of the most extensively studied research topics within the field of control engineering. This design process usually breaks into two distinct parts: in the first part the switching surface is selected, and in the second one the control signal which always makes the system representative point approach the surface is chosen. Once the representative point hits the surface, then under the same control signal, the point remains on the surface. Thus, the switching surface fully determines the system dynamics in the sliding mode and should be carefully selected by the system designer.

Therefore, the main topic of this book is the issue of the switching surface design. We considered two standard plants, which are very often encountered in the control engineering practice. These are the second and the third order nonlinear and possibly time-varying systems. For these two systems we propose sliding mode control without the reaching phase, which is possible due to the application of the time-varying sliding lines for the second order systems and the time-varying sliding planes for the third order systems. The switching surfaces introduced in this book are chosen in such a way that system representative point belongs to them at the beginning of the regulation process. Then, the surfaces move smoothly in the phase space to their predetermined fixed position. By this means the reaching phase is eliminated and the systems – unlike in conventional variable structure control with the fixed, time-invariant switching surfaces – become insensitive with respect to matched external disturbance and modelling uncertainty from the very beginning of the regulation process. However, the main contribution of this monograph is not only the idea of using time-varying switching surfaces, but more importantly the set of design procedures which generate the feasible switching lines and planes for the controlled systems subject to various constraints. The constraints taken into account are fairly natural and they comprise input signal and velocity limits for second order systems, and additionally input signal, velocity and acceleration thresholds for third order systems. The design procedures described in this book generate the switching lines and planes which ensure the system performance optimal in the sense of integral absolute error and also in the sense of integral time multiplied absolute error. Both conventional constraints described by inequalities and more practical, elastic constraints expressed by appropriate penalty functions introduced into the performance indices are taken into account in the design process.

A. Bartoszewicz and A. Nowacka-Leverton: Time-Varying Sliding Modes, LNCIS 382, pp. 181–182.
springerlink.com   © Springer-Verlag Berlin Heidelberg 2009

The results presented in this monograph are to some extent related to the methodology, usually called the reaching law approach which is popular in the sliding mode control literature. In that approach, the desired evolution of the switching variable is specified first, and then a control law ensuring that this variable changes according to the specification is determined. The approach is in a way similar to the application of the time-varying switching surfaces in the control systems. However, we believe that the reaching law approach, in its present form published in many outstanding papers on this subject, suffers from some drawback, namely a lack of relation between the physical constraints of the controlled system and the reaching law specifying the desired system performance. Therefore, the controller design methods proposed in this monograph may be seen as the first step (or the first two steps, since we consider the second and the third order systems) along the path to designing reaching laws which will explicitly take into account physical limitations of the controlled plants. Indeed, the design methods presented in this book combine the reaching law selection with the choice of the fixed, time-invariant switching surfaces. Consequently, they offer better results than previous methods which perform the two tasks separately.

Furthermore, we believe that the work presented in this book might possibly be extended by considering several other classes (or families) of the time varying switching surfaces. These could include terminal attractors, i.e. the switching surfaces ensuring not only asymptotic but also finite time error convergence, and other time varying nonlinear switching surfaces which might offer better dynamic performance of the system with the same input signal and state constraints. The design methods proposed in this book can also be generalised for the higher order continuous and discrete time sliding mode control systems. Furthermore, the application of observers in order to avoid the need of all state variable measurement and obtain output feedback controllers is also a very promising research direction.

References

Ambrosino, G., Celentano, G., Garofalo, F.: Variable structure model reference adaptive control systems. Int. Journal of Control 39, 1339–1349 (1984)

Bandyopadhyay, B., Janardhanan, S.: Discrete-time Sliding Mode Control: A Multirate-Output Feedback Approach. Lecture Notes in Control and Information Sciences, vol. 323. Springer, Heidelberg (2005)

Bartolini, G., Ferrara, A., Spurgeon, S. (eds.): New trends in sliding mode control. Special issue: Int. Journal of Robust and Nonlinear Control 7, 297–427 (1997a)

Bartolini, G., Ferrara, A., Usai, E.: Applications of a sub optimal discontinuous control algorithm for uncertain second order systems. Int. Journal of Robust and Nonlinear Control 7, 299–319 (1997b)

Bartolini, G., Ferrara, A., Usai, E.: Output tracking control of uncertain non linear second order systems. Automatica 33, 2203–2212 (1997c)

Bartolini, G., Ferrara, A., Usai, E.: Chattering avoidance by second-order sliding mode control. IEEE Transactions on Automatic Control 43, 241–246 (1998)

Bartolini, G., Fridman, L., Pisano, A. (eds.): Modern sliding mode control theory. New perspectives and applications. Lecture Notes in Control and Information Sciences, vol. 375. Springer, Heidelberg (2008)

Bartolini, G., Pydynowski, P.: An improved, chattering free, VSC scheme for uncertain dynamical systems. IEEE Transactions on Automatic Control 41, 1220–1226 (1996)

Bartoszewicz, A.: A comment on a time-varying sliding surface for fast and robust tracking control of second-order uncertain systems. Automatica 31, 1893–1895 (1995)

Bartoszewicz, A.: Time-varying sliding modes for second-order systems. Proc. of the IEE – Part D: Control Theory and Applications 143, 455–462 (1996)

Bartoszewicz, A.: Discrete time quasi-sliding mode control strategies. IEEE Transactions on Industrial Electronics 45, 633–637 (1998)

Bartoszewicz, A.: Chattering attenuation in sliding mode control systems. Control and Cybernetics 29, 585–594 (2000)

Bartoszewicz, A., Kaynak, O., Utkin, V. (eds.): Sliding mode control in industrial applications. Special section: IEEE Transactions on Industrial Electronics, vol. 55, 3806–4074 (2008)

Bartoszewicz, A., Nowacka, A.: A time-varying switching plane design for variable structure control of the third order system with input constraint. Archives of Control Sciences 14, 201–213 (2004a)

Bartoszewicz, A., Nowacka, A.: Sliding mode control of the third order system with input constraint. In: Proc. of the 10th IEEE Conference on Methods and Models in Automation and Robotics, vol. 1, pp. 305–310 (2004b)

Bartoszewicz, A., Nowacka, A.: A moving switching plane for the sliding mode control of the third order system. In: Proc. of the 16th IFAC World Congress (2005a)

Bartoszewicz, A., Nowacka, A.: Sliding mode control of the third order system with state constraints. Sys. Science 30, 35–44 (2005b)

Bartoszewicz, A., Nowacka, A.: Switching plane design for the sliding mode control of systems with elastic input constraints. Proc. of the Institution of Mechanical Eng. Part I – Journal of Sys. and Control Eng. 219, 393–403 (2005c)

Bartoszewicz, A., Nowacka, A.: An optimal switching plane design for the third order system subject to state constraints. In: Proc. of the 9th Int. Workshop on Variable Structure Sys, pp. 86–91 (2006a)

Bartoszewicz, A., Nowacka, A.: Optimal design of the shifted switching planes for VSC of the third order system. The Transactions of the Inst. of Measurement and Control 28, 335–352 (2006b)

Bartoszewicz, A., Nowacka, A.: Reaching phase elimination in variable structure control of the third order system with state constraints. Kybernetika (Cybernetics) 42, 111–126 (2006c)

Bartoszewicz, A., Nowacka, A.: Shifted switching plane design for the third order systems with elastic and conventional input constraints. Circuits Sys. and Signal Processing 25, 661–684 (2006d)

Bartoszewicz, A., Nowacka, A.: Sliding mode control of the third order system subject to velocity acceleration and input signal constraints. Int. Journal of Adaptive Control and Signal Processing 21, 779–794 (2007a)

Bartoszewicz, A., Nowacka, A.: VSC of the third order systems with state and input signal constraints. In: Proc. of the European Control Conference, pp. 3203–3210 (2007b)

Bartoszewicz, A., Nowacka-Leverton, A.: SMC without the reaching phase – the switching plane design for the third order system. Proc. of the IET – Part D: Control Theory and Applications 1, 1461–1470 (2007)

Bartoszewicz, A., Patton, R.J. (eds.): Sliding mode control. Special issue: Int. Journal of Adaptive Control and Signal Processing 21, 635–822 (2007)

Bartoszewicz, A., Żuk, J.: Time-varying switching lines for VSC of robot manipulators. Transactions of the VSB – Technical University of Ostrava LII, 13–18 (2006)

Betin, F., Pinchon, D., Capolino, G.: A time-varying sliding surface for robust position control of a DC motor drive. IEEE Transactions on Industrial Electronics 49, 462–473 (2002a)

Betin, F., Sivert, A., Pinchon, D.: Time-varying sliding modes for a DC motor drive. In: Proc. of the Int. Symposium on Industrial Electronics, pp. 378–382 (2002b)

Bhatti, A.: Advanced Sliding Mode Controllers for Industrial Applications – PhD dissertation. University of Leicester, Leicester (1998)

Bhatti, A., Spurgeon, S., Dorey, R., Edwards, C.: Sliding mode configurations for automotive engine control. Int. Journal of Adaptive Control and Signal Processing 13, 49–69 (1999)

Burton, J.A., Zinober, A.: Continuous approximation of variable structure control. Int. Journal of Sys. Science 17, 875–885 (1986)

Chakravarthini Saaj, M., Bandyopadhyay, B.: Discrete output feedback sliding mode control – a moving switching surface approach. Sys. Science 27, 5–21 (2001)

Chang, T.H., Hurmuzlu, Y.: Trajectory tracking in robotic systems using variable structure control without a reaching phase. In: Proc. of the American Control Conference, pp. 1505–1509 (1992)

Chang, T.H., Hurmuzlu, Y.: Sliding control without reaching phase and its application to bipedal locomotion. Transactions of the ASME – Journal of Dyn. Sys., Measurement, and Control 115, 447–455 (1993)

References

Chen, Z.M., Zhang, J.G., Zhao, Z.C., Zeng, J.C.: A new method of fuzzy time-varying sliding mode control. In: Proc. of the First Int. Conference on Machine Learning and Cybernetics, pp. 1789–1792 (2002)

Choi, S.B., Cheong, C.C., Park, D.W.: Moving switching surfaces for robust control of second-order variable structure systems. Int. Journal of Control 58, 229–245 (1993)

Choi, S.B., Kim, J.S.: A fuzzy-sliding mode controller for robust tracking of robotic manipulators. Mechatronics 7, 199–216 (1997)

Choi, S.B., Park, D.W.: Moving sliding surfaces for fast tracking control of second-order dynamic systems. Transactions of the ASME – Journal of Dyn. Sys., Measurement, and Control 116, 154–158 (1994)

Choi, S.B., Park, D.W., Jayasuriya, S.: A time-varying sliding surface for fast and robust tracking control of second-order uncertain systems. Automatica 30, 899–904 (1994)

Chung, C.C., Jang, J.W., Lee, H.S.: Design of seek control of hard disc drive using discrete-time sliding mode control. In: Proc. of the IEEE Int. Conference on Control Applications, pp. 533–538 (2004)

Chung, S.C., Lin, C.L.: A transformed Luré problem for sliding mode control and chattering reduction. IEEE Transactions on Automatic Control 44, 563–568 (1999)

Corradini, M.L., Orlando, G.: Linear unstable plants with saturating actuators: robust stabilization by a time varying sliding surface. Automatica 43, 88–94 (2007)

DeCarlo, R., Żak, S., Matthews, G.: Variable structure control of nonlinear multivariable systems: a tutorial. Proc. of IEEE 76, 212–232 (1988)

Demin, X., Weisheng, Y., Yang, S.: Nonlinear variable structure double mode control of AUV. In: Proc. of the Int. Symposium on Underwater Technology, pp. 425–430 (2000)

Draženović, B.: The invariance conditions in variable structure systems. Automatica 5, 287–295 (1969)

Edwards, C., Spurgeon, S.: Sliding mode output tracking with application to a multivariable high temperature furnace problem. Int. Journal of Robust and Nonlinear Control 7, 337–351 (1997)

Edwards, C., Spurgeon, S.: Sliding mode control: theory and applications. Taylor & Francis, London (1998)

Edwards, C., Fossas Colet, E., Fridman, L. (eds.): Advances in variable structure and sliding mode control. Lecture Notes in Control and Information Sciences, vol. 334. Springer, Heidelberg (2006)

Elmali, H., Olgac, N.: Implementation of sliding mode control with perturbation estimation (SMCPE). IEEE Transactions on Control Sys. Technology 4, 79–85 (1996)

Емельянов, С.В. (ed.): Теория Систем с Переменной Структурой. Издателство Наука, Москва (1970)

Gao, W. (ed.): Variable structure control. Special section: IEEE Transactions on Industrial Electronics 40, 1–88 (1993)

Gao, W., Wang, Y., Homaifa, A.: Discrete-time variable structure control systems. IEEE Transactions on Industrial Electronics 42, 117–122 (1995)

Ghosh, R.R., Olgac, N.: Robust nonlinear control via moving sliding surfaces – n-th order case. In: Proc. of the 36th IEEE Conference on Decis. and Control, pp. 943–948 (1997)

Ha, Q., Rye, D., Durrant-Whyte, H.: Fuzzy moving sliding mode control with application to robotic manipulators. Automatica 35, 607–616 (1999)

Hara, M., Furuta, K., Pan, Y., Hoshino, T.: Evaluation of discrete-time VSC on an inverted pendulum apparatus with additional dynamics. Appl. Mathematics and Computer Science 8, 159–181 (1998)

Hung, J.Y., Gao, W., Hung, J.C.: Variable structure control: a survey. IEEE Transactions on Industrial Electronics 40, 2–22 (1993)

Ignaciuk, P., Bartoszewicz, A.: Linear quadratic optimal discrete time sliding mode controller for connection-oriented communication networks. IEEE Transactions on Industrial Electronics 55, 4013–4021 (2008a)

Ignaciuk, P., Bartoszewicz, A.: Linear quadratic optimal sliding mode flow control for connection-oriented communication networks. Int. Journal of Robust and Nonlinear Control 18 (2008b)

Iliev, B., Hristozov, I.: Variable structure control using Takagi-Sugeno fuzzy system as a sliding surface. In: Proc. of the IEEE Int. Conference on Fuzzy Sys., pp. 644–649 (2002)

Iliev, B., Kalaykov, I.: Improved sliding mode robot control – a fuzzy approach. In: Proc. of the 3rd Int. Workshop on Robot Motion and Control, pp. 393–398 (2002)

Iliev, B., Kalaykov, I.: Minimum-time sliding mode control for second-order systems. In: Proc. of the American Control Conference, pp. 626–631 (2004)

Itkis, U.: Control systems of variable structure. Wiley, New York (1976)

Janardhanan, S., Kariwala, V.: Multirate-output-feedback-based LQ-optimal discrete-time sliding mode control. IEEE Transactions on Automatic Control 53, 367–373 (2008)

Jezernik, K.: VSS control of unity power factor. IEEE Transactions on Industrial Electronics 46, 325–332 (1999)

Jezernik, K., Rutar, D., Milanovic, M.: Nonlinear voltage control of unity power factor boost converter. In: Proc. of the IEEE Int. Symposium on Industrial Electronics, pp. 492–497 (1993)

Jinggang, Z., Jiang, K., Chen, Z., Zhao, Z.: Global robust fuzzy sliding mode control for a class of non-linear system. Transactions of the Inst. of Measurement and Control 28, 219–227 (2006)

Jinggang, Z., Yibo, Z., Zhimei, C.: A discrete variable structure control scheme based on combined switching surfaces for position control systems. In: Proc. of the 4th Int. Power Electronics and Motion Control Conference, pp. 738–741 (2004a)

Jinggang, Z., Yibo, Z., Zhimei, C., Zhicheng, Z.: A control scheme based on discrete time-varying sliding surface for position control systems. In: Proc. of the 5th World Congress on Intelligent Control and Automation, pp. 1175–1178 (2004b)

Kaynak, O. (ed.): Computationally intelligent methodologies and sliding-mode control. Special section: IEEE Transactions on Industrial Electronics 48, 2–240 (2001)

Kim, N.I., Lee, C.W., Chang, P.H.: Sliding mode control with perturbation estimation: application to motion control of parallel manipulator. Control Eng. Practice 6, 1321–1330 (1998)

Lai, Y.M., Hwang, R.C., Chen, C.J., Chuang, C.W., Yu, G.R.: A new fuzzy reaching law for discrete-time variable structure control systems. In: Proc. of Joint 9th IFSA World Congress and 20th NAFIPS Int. Conference, pp. 2803–2806 (2001)

Levant, A.: Sliding order and sliding accuracy in sliding mode control. Int. Journal of Control 58, 1247–1263 (1993)

Levant, A.: Higher-order sliding modes, differentiation and output-feedback control. Int. Journal of Control 76, 924–941 (2003)

Lu, X.Y., Spurgeon, S.: A new sliding mode approach to asymptotic feedback linearisation with application to the control of non-flat systems. Appl. Mathematics and Computer Science 8, 21–37 (1998)

References

Ma, Y., Hong, G.S., Poo, A.N.: Ideal motion control using pre-shape sliding mode controller. In: Proc. of the IEEE Int. Conference on Robotics and Automation, pp. 2159–2164 (1999)

Misawa, E., Utkin, V. (eds.): Variable structure systems. Special issue: Transactions of the ASME – Journal of Dyn. Sys., Measurement, and Control 122, 585–819 (2000)

Nowacka, A., Bartoszewicz, A.: Sliding mode control of the third order system with input and acceleration constraint. In: Proc. of the 11th IEEE Conference on Methods and Models in Automation and Robotics, pp. 373–378 (2005)

Nowacka, A., Bartoszewicz, A.: A new procedure of time-varying switching plane design. In: Proc. of the 32nd Annual Conference of the IEEE Industrial Electronics Society, pp. 543–548 (2006a)

Nowacka, A., Bartoszewicz, A.: Moving switching plane selection for the third order system subject to input and acceleration constraints. In: Proc. of the 32nd Annual Conference of the IEEE Industrial Electronics Society, pp. 671–676 (2006b)

Nowacka, A., Bartoszewicz, A.: Sliding mode control of the third order system – the optimal selection of the switching plane. In: Proc. of the 12th IEEE Conference on Methods and Models in Automation and Robotics, pp. 77–84 (2006c)

Nowacka, A., Bartoszewicz, A.: Stretchable and conventional input constraints in VSC of the third order systems. In: Proc. of the 9th Int. Workshop on Variable Structure Sys., pp. 80–85 (2006d)

Nowacka-Leverton, A., Bartoszewicz, A.: IAE optimal sliding mode control algorithm for second order systems with velocity and input signal constraints. In: Proc. of the 8th Int. Conference on Technical Informatics, pp. 5–12 (2008a)

Nowacka-Leverton, A., Bartoszewicz, A.: ITAE optimal SMC method for second order systems with velocity and input signal constraints. In: Proc. of the 14th Int. Congress of Cybernetics and Sys. of WOSC, pp. 425–435 (2008b)

Nowacka-Leverton, A., Bartoszewicz, A.: SMC of second order systems subject to elastic constraints. In: Int. Carpathian Control Conference, pp. 454–457 (2008c)

Palm, R.: Robust control by fuzzy sliding mode. Automatica 30, 1429–1437 (1994)

Palm, R., Driankov, D., Hellendoorn, H.: Model based fuzzy control. Springer, Berlin (1997)

Pan, Y., Furuta, K.: Variable structure control with sliding sector based on hybrid switching law. Int. Journal of Adaptive Control and Signal Processing 21, 764–778 (2007)

Park, D.W., Choi, S.B.: Moving sliding surfaces for high – order variable structure systems. Int. Journal of Control 72, 960–970 (1999)

Park, C.W., Kim, J.H., Kwon, C., Park, M.: Tracking control of a robot manipulator using sliding mode controller with fast and accurate performance. In: Proc. of the IEEE/RSJ Int. Conference on Intelligent Robots and Sys., pp. 305–310 (1999)

Rios-Bolivar, M., Zinober, A., Sira-Ramirez, H.: Dynamical adaptive sliding mode output tracking control of a class of nonlinear systems. Int. Journal of Robust and Nonlinear Control 7, 387–405 (1997)

Sabanovic, A., Fridman, L., Spurgeon, S.: Variable structure systems: from principles to implementation. IEE Book Series, London (2004)

Selisteanu, D., Petre, E., Rasvan, V.B.: Sliding mode and adaptive sliding-mode control of a class of nonlinear bioprocesses. Int. Journal of Adaptive Control and Signal Processing 21, 795–822 (2007)

Shtessel, Y., Fridman, L., Zinober, A.: Special issue: Advances in higher order sliding mode control. Int. Journal of Robust and Nonlinear Control 18, 381–585 (2008)

Shtessel, Y., Lee, Y.J.: New approach to chattering analysis in systems with sliding modes. In: Proc. of the 35th Conference on Decis. and Control, pp. 4014–4019 (1996)

Shyu, K., Tsai, Y., Yung, C.: A modified variable structure controller. Automatica 28, 1209–1213 (1992)

Sira-Ramirez, H.: A dynamical variable structure control strategy in asymptotic output tracking problems. IEEE Transactions on Automatic Control 38, 615–620 (1993a)

Sira-Ramirez, H.: On the dynamical sliding mode control of nonlinear systems. Int. Journal of Control 57, 1039–1061 (1993b)

Sira-Ramirez, H., Llanes-Santiago, O.: Dynamical discontinuous feedback strategies in the regulation of nonlinear chemical processes. IEEE Transactions on Control Sys. Technology 2, 11–21 (1994)

Sivert, A., Betin, F., Faqir, A., Capolino, G.A.: Robust control of an induction machine drive using a time-varying sliding surface. In: Proc. of the IEEE Int. Symposium on Industrial Electronics, pp. 1369–1374 (2004a)

Sivert, A., Betin, F., Faqir, A., Capolino, G.A.: Time-varying sliding surface for position control of an induction machine drive. In: Proc. of the 16th Int. Conference on Electr. Machines (2004b)

Sivert, A., Faqir, A., Nahidmobarakeh, B., Betin, F., Capolino, G.A.: Moving switching surfaces for high precision control of electrical drives. In: Proc. of the IEEE Int. Conference on Industrial Technology, pp. 175–180 (2004c)

Slotine, J., Li, W.: Applied nonlinear control. Prentice-Hall Int., Englewood Cliffs (1991)

Spong, M.W., Vidyasagar, M.: Robot dynamics and control. Wiley, New York (1989)

Spurgeon, S.K., Davies, R.A.: Nonlinear control strategy for robust sliding mode performance. Int. J. Control 57, 1107–1123 (1993)

Spurgeon, S.K., Lu, X.Y.: Output tracking using dynamic sliding mode techniques. Int. Journal of Robust and Nonlinear Control 7, 407–427 (1997)

Temeltas, H.: A fuzzy adaptation technique for sliding mode controllers. In: Proc. of the IEEE Int. Symposium on Industrial Electronics, pp. 110–115 (1998)

Tokat, S., Eksin, I., Guzelkaya, M.: A new design method for sliding mode controllers using a linear time-varying sliding surface. Proc. of the Institution of Mechanical Eng. Part I – Journal of Sys. and Control Eng. 216, 455–466 (2002)

Tokat, S., Eksin, I., Guzelkaya, M.: New approaches for on-line tuning of the linear sliding surface slope in sliding mode controllers. Turkish Journal of Electr. Eng. and Computer Science 11, 45–59 (2003a)

Tokat, S., Eksin, I., Guzelkaya, M., Soylemez, M.T.: Design of a sliding mode controller with a nonlinear time-varying sliding surface. Transactions of the Inst. of Measurement and Control 25, 145–162 (2003b)

Уткин, В.И.: Об уравнениях скользящего режима в разрывных системах I. Автоматика и Телемеханика 32, 42–54 (1971)

Уткин, В.И.: Об уравнениях скользящего режима в разрывных системах II. Автоматика и Телемеханика 33, 51–61 (1972)

Уткин, В.И.: Скользящие Режимы и их Применения в Системах с Переменной Структурой. Издательство Наука, Москва (1974)

Utkin, V.: Variable structure systems with sliding modes. IEEE Transactions on Automatic Control 22, 212–222 (1977)

Utkin, V.: Sliding Modes in Control and Optimization Springer-Verlag, Berlin (1992); also Уткин, В.И.: Скользящие Режимы в Задачах Оптимизации и Управления. Издательство Наука, Москва (1981)

References 189

Utkin, V. (ed.): Sliding mode control. Special issue: Int. Journal of Control 57, 1003–1259 (1993)

Utkin, V.: Sliding mode control design principles and applications to electric drives. IEEE Transactions on Industrial Electronics 40, 23–36 (1993)

Utkin, V., Guldner, J., Shi, J.: Sliding mode control in electromechanical systems. Taylor & Francis, London (1999)

Weisheng, Y., Demin, X., Zhang, R.: Global sliding-mode control for companion nonlinear system with bounded control. In: Proc. of the American Control Conference, pp. 3884–3888 (1998)

Xu, J., Lee, T., Wang, M., Yu, X.: Design of variable structure controllers with continuous switching control. Int. Journal of Control 65, 409–431 (1996)

Yager, R.R., Filev, D.P.: Essentials of fuzzy modeling and control. Wiley, New York (1994)

Yagiz, N., Hacioglu, Y.: Fuzzy sliding modes with moving surface for the robust control of a planar robot. Journal of Vib. and Control 11, 903–922 (2005)

Yilmaz, C., Hurmuzlu, Y.: Eliminating the reaching phase from variable structure control. Transactions of the ASME – Journal of Dyn. Sys., Measurement, and Control 122, 753–757 (2000)

Yu, X.H. (ed.): Adaptive learning and control using sliding modes. Special issue: Appl. Mathematics and Computer Science 8, 5–197 (1998)

Yu, H., Lloyd, S.: Variable structure adaptive control of robot manipulators. Proc. of the IEE – Part D: Control Theory and Applications 144, 167–176 (1997)

Zhang, D., Guo, G.: Discrete-time sliding mode proximate time optimal seek control of hard disk drives. Proc. of the IEE – Part D: Control Theory and Applications 147, 440–446 (2000)

Zhiming, J., Shengwei, W., Tingqi, L.: Variable structure control of electrohydraulic servo systems using fuzzy CMAS neural network. Transactions of the Inst. of Measurement and Control 25, 185–201 (2003)

Zinober, A. (ed.): Variable Structure and Lyapunov Control. Springer, London (1994)

Zlateva, P.: Variable-structure control of nonlinear systems. Control Eng. Practice 4, 1023–1028 (1996)

Index

Accelaration constraint, *see* constraint, acceleration
asymptotic error convergence, *see* error convergence, asymptotic
autonomous underwater vehicle, 12

bisection, 112
boundary layer controller, 9

Chattering, 7, 9, 11, 12
computed torque method, 12
constraint, 11, 13
   acceleration, 86, 88, 89, 96, 99, 149
   elastic, 16, 22, 42, 52, 181
   elastic acceleration, 90, 151
   elastic input signal, 31, 52, 148
   elastic velocity, 35, 37, 38, 57, 59, 101, 103, 107, 159, 161
   input signal, 14, 24, 27, 42, 47, 73, 86, 120, 140, 168, 173,
   velocity, 33, 34, 38, 39, 56, 57, 58, 59, 99, 100, 101, 104, 105, 107, 138, 140, 156, 158
control signal, 4, 11, 12, 13, 15, 26, 38, 48, 77, 99, 107, 120, 138, 181

Dc motor, 10, 11, 13
demand trajectory, *see* trajectory, demand
describing function, 9
desired trajectory, *see* trajectory, desired
discontinuous feedback, 1
discrete time systems, 10, 12

Elastic acceleration constraint, see constraint , elastic acceleration
elastic constraint, *see* constraint, elastic
elastic velocity constraint, *see* constraint, elastic velocity

error convergence
   asymptotic, 149
   finite time, 3, 10, 182
external disturbance, 6, 10, 13, 14, 18, 22, 30, 37, 52, 59, 86, 95, 99, 107, 112, 115, 120, 138, 139, 181

Finite time error convergence, *see* error convergence, finite time

Hard disk drive, 12
hypersurface intersection, 6

IAE, 13, 22, 23, 26, 48, 52, 57, 58, 59, 64, 65, 72, 77, 138, 139, 158, 165, 168, 169, 173
induction motor, 11
initial conditions, 4, 6, 13, 17, 20, 21, 30, 37, 52, 59, 69, 70, 86, 95, 107, 112, 115, 120, 138
input signal constraint, *see* constraint, input signal
ITAE, 13, 22, 46, 48, 51, 52, 56, 58, 59, 62, 64, 65, 139, 148, 149, 158, 159, 161, 165, 168, 169, 173

Lyapunov stability, 7

Matched disturbance, 7
model uncertainty, 1, 6, 7, 10, 18, 22, 30, 52, 86, 95, 99, 107, 112, 115, 117, 120, 138, 139
multi-input system, 6
multirate sampling, 13

Order reduction, 6

Penalty function, 26, 48, 77
phase
    plane, 7, 10, 13, 18
    space, 12, 181
    trajectory, 4, 31, 38
point to point (PTP) control, 15, 18

Reaching
    law, 7, 13, 182
    phase, 6, 9, 11, 13, 18, 19, 181
representative point, 4, 13, 15, 19, 68, 181
robot manipulator, 11, 14, 15, 18

Second order system, 2, 12, 17, 18, 30, 44, 181
sliding
    hypersurface, 10, 12
    line, 6, 10, 13, 27, 31, 49, 181
    line slope, 6
    mode control, 1, 4, 9, 10, 11, 12, 13, 16, 18, 181, 182
    mode system, 1
    mode technique, 1
    phase, 6
    surface, 10, 67
state
    variables, 2, 7, 13, 17, 67
    vector, 2, 17, 67
switching
    hypersurface, 6
    line, 6, 10, 11, 13, 14, 18, 19, 20, 21, 22, 26, 29, 30, 35, 37, 42, 44, 47, 48, 57, 59, 62, 64
    line parameters, 6, 22, 23, 24, 26, 29, 42, 44, 46, 47, 51, 52, 57, 64
    line slope, 48

plane, 13, 15, 68, 69, 70, 71, 72, 77, 83, 86, 87, 89, 90, 94, 95, 99, 112, 117, 123, 130, 138, 139, 140, 142, 143, 148, 149, 150, 155, 160, 161, 165, 169, 174
plane parameters, 72, 77, 83, 86, 89, 94, 95, 99, 123, 138, 140, 142, 148, 155, 161, 165, 174
strategy, 2, 3, 4
surface, 6, 181
surface intersection, 6
variable, 7, 8, 182
system
    insensitivity, 7, 13, 18
    uncertainty, 17, 18, 67

Terminal attractors, 182
third order system, 15, 67, 68, 95, 107, 138, 139, 181, 182
tracking error, 14, 15, 17, 18, 20, 22, 23, 25, 26, 30, 38, 46, 52, 67, 69, 70, 71, 73, 75, 86, 96, 97, 98, 99, 104, 105, 106, 107, 112, 116, 117, 120, 121, 138, 139, 143
trajectory
    demand, 12, 30, 37, 52, 59, 86, 98, 107, 112, 115, 120, 138
    desired, 17, 67
twisting controller, 4

Variable structure
    control, 3, 10
    systems, 1
velocity constraint, *see* constraint, velocity

Lecture Notes in Control and Information Sciences

Edited by M. Thoma, F. Allgöwer, M. Morari

Further volumes of this series can be found on our homepage:
springer.com

Vol. 382: Bartoszewicz A.;
Nowacka-Leverton A.;
Time-Varying Sliding Modes for Second
and Third Order Systems
192 p. 2009 [978-3-540-92216-2]

Vol. 381: Hirsch M.J.; Commander C.W.;
Pardalos P.M.; Murphey R. (Eds.)
Optimization and Cooperative Control Strategies:
Proceedings of the 8th International Conference
on Cooperative Control and Optimization
459 p. 2009 [978-3-540-88062-2]

Vol. 380: Basin M.
New Trends in Optimal Filtering and Control for
Polynomial and Time-Delay Systems
206 p. 2008 [978-3-540-70802-5]

Vol. 379: Mellodge P.; Kachroo P.;
Model Abstraction in Dynamical Systems:
Application to Mobile Robot Control
116 p. 2008 [978-3-540-70792-9]

Vol. 378: Femat R.; Solis-Perales G.;
Robust Synchronization of Chaotic Systems
Via Feedback
199 p. 2008 [978-3-540-69306-2]

Vol. 377: Patan K.
Artificial Neural Networks for
the Modelling and Fault
Diagnosis of Technical Processes
206 p. 2008 [978-3-540-79871-2]

Vol. 376: Hasegawa Y.
Approximate and Noisy Realization of
Discrete-Time Dynamical Systems
245 p. 2008 [978-3-540-79433-2]

Vol. 375: Bartolini G.; Fridman L.; Pisano A.;
Usai E. (Eds.)
Modern Sliding Mode Control Theory
465 p. 2008 [978-3-540-79015-0]

Vol. 374: Huang B.; Kadali R.
Dynamic Modeling, Predictive Control
and Performance Monitoring
240 p. 2008 [978-1-84800-232-6]

Vol. 373: Wang Q.-G.; Ye Z.; Cai W.-J.;
Hang C.-C.
PID Control for Multivariable Processes
264 p. 2008 [978-3-540-78481-4]

Vol. 372: Zhou J.; Wen C.
Adaptive Backstepping Control of Uncertain
Systems
241 p. 2008 [978-3-540-77806-6]

Vol. 371: Blondel V.D.; Boyd S.P.;
Kimura H. (Eds.)
Recent Advances in Learning and Control
279 p. 2008 [978-1-84800-154-1]

Vol. 370: Lee S.; Suh I.H.;
Kim M.S. (Eds.)
Recent Progress in Robotics:
Viable Robotic Service to Human
410 p. 2008 [978-3-540-76728-2]

Vol. 369: Hirsch M.J.; Pardalos P.M.;
Murphey R.; Grundel D.
Advances in Cooperative Control and
Optimization
423 p. 2007 [978-3-540-74354-5]

Vol. 368: Chee F.; Fernando T.
Closed-Loop Control of Blood Glucose
157 p. 2007 [978-3-540-74030-8]

Vol. 367: Turner M.C.; Bates D.G. (Eds.)
Mathematical Methods for Robust and Nonlinear
Control
444 p. 2007 [978-1-84800-024-7]

Vol. 366: Bullo F.; Fujimoto K. (Eds.)
Lagrangian and Hamiltonian Methods for
Nonlinear Control 2006
398 p. 2007 [978-3-540-73889-3]

Vol. 365: Bates D.; Hagström M. (Eds.)
Nonlinear Analysis and Synthesis Techniques for
Aircraft Control
360 p. 2007 [978-3-540-73718-6]

Vol. 364: Chiuso A.; Ferrante A.;
Pinzoni S. (Eds.)
Modeling, Estimation and Control
356 p. 2007 [978-3-540-73569-4]

Vol. 363: Besançon G. (Ed.)
Nonlinear Observers and Applications
224 p. 2007 [978-3-540-73502-1]

Vol. 362: Tarn T.-J.; Chen S.-B.;
Zhou C. (Eds.)
Robotic Welding, Intelligence and Automation
562 p. 2007 [978-3-540-73373-7]

Vol. 361: Méndez-Acosta H.O.; Femat R.; González-Álvarez V. (Eds.):
Selected Topics in Dynamics and Control of Chemical and Biological Processes
320 p. 2007 [978-3-540-73187-0]

Vol. 360: Kozlowski K. (Ed.)
Robot Motion and Control 2007
452 p. 2007 [978-1-84628-973-6]

Vol. 359: Christophersen F.J.
Optimal Control of Constrained Piecewise Affine Systems
190 p. 2007 [978-3-540-72700-2]

Vol. 358: Findeisen R.; Allgöwer F.; Biegler L.T. (Eds.): Assessment and Future Directions of Nonlinear Model Predictive Control
642 p. 2007 [978-3-540-72698-2]

Vol. 357: Queinnec I.; Tarbouriech S.; Garcia G.; Niculescu S.-I. (Eds.):
Biology and Control Theory: Current Challenges
589 p. 2007 [978-3-540-71987-8]

Vol. 356: Karatkevich A.:
Dynamic Analysis of Petri Net-Based Discrete Systems
166 p. 2007 [978-3-540-71464-4]

Vol. 355: Zhang H.; Xie L.:
Control and Estimation of Systems with Input/Output Delays
213 p. 2007 [978-3-540-71118-6]

Vol. 354: Witczak M.:
Modelling and Estimation Strategies for Fault Diagnosis of Non-Linear Systems
215 p. 2007 [978-3-540-71114-8]

Vol. 353: Bonivento C.; Isidori A.; Marconi L.; Rossi C. (Eds.)
Advances in Control Theory and Applications
305 p. 2007 [978-3-540-70700-4]

Vol. 352: Chiasson, J.; Loiseau, J.J. (Eds.)
Applications of Time Delay Systems
358 p. 2007 [978-3-540-49555-0]

Vol. 351: Lin, C.; Wang, Q.-G.; Lee, T.H., He, Y.
LMI Approach to Analysis and Control of Takagi-Sugeno Fuzzy Systems with Time Delay
204 p. 2007 [978-3-540-49552-9]

Vol. 350: Bandyopadhyay, B.; Manjunath, T.C.; Umapathy, M.
Modeling, Control and Implementation of Smart Structures 250 p. 2007 [978-3-540-48393-9]

Vol. 349: Rogers, E.T.A.; Galkowski, K.; Owens, D.H.
Control Systems Theory and Applications for Linear Repetitive Processes
482 p. 2007 [978-3-540-42663-9]

Vol. 347: Assawinchaichote, W.; Nguang, K.S.; Shi P.
Fuzzy Control and Filter Design for Uncertain Fuzzy Systems
188 p. 2006 [978-3-540-37011-6]

Vol. 346: Tarbouriech, S.; Garcia, G.; Glattfelder, A.H. (Eds.)
Advanced Strategies in Control Systems with Input and Output Constraints
480 p. 2006 [978-3-540-37009-3]

Vol. 345: Huang, D.-S.; Li, K.; Irwin, G.W. (Eds.)
Intelligent Computing in Signal Processing and Pattern Recognition
1179 p. 2006 [978-3-540-37257-8]

Vol. 344: Huang, D.-S.; Li, K.; Irwin, G.W. (Eds.)
Intelligent Control and Automation
1121 p. 2006 [978-3-540-37255-4]

Vol. 341: Commault, C.; Marchand, N. (Eds.)
Positive Systems
448 p. 2006 [978-3-540-34771-2]

Vol. 340: Diehl, M.; Mombaur, K. (Eds.)
Fast Motions in Biomechanics and Robotics
500 p. 2006 [978-3-540-36118-3]

Vol. 339: Alamir, M.
Stabilization of Nonlinear Systems Using Receding-horizon Control Schemes
325 p. 2006 [978-1-84628-470-0]

Vol. 338: Tokarzewski, J.
Finite Zeros in Discrete Time Control Systems
325 p. 2006 [978-3-540-33464-4]

Vol. 337: Blom, H.; Lygeros, J. (Eds.)
Stochastic Hybrid Systems
395 p. 2006 [978-3-540-33466-8]

Vol. 336: Pettersen, K.Y.; Gravdahl, J.T.; Nijmeijer, H. (Eds.)
Group Coordination and Cooperative Control
310 p. 2006 [978-3-540-33468-2]

Vol. 335: Kozłowski, K. (Ed.)
Robot Motion and Control
424 p. 2006 [978-1-84628-404-5]

Vol. 334: Edwards, C.; Fossas Colet, E.; Fridman, L. (Eds.)
Advances in Variable Structure and Sliding Mode Control
504 p. 2006 [978-3-540-32800-1]

Vol. 333: Banavar, R.N.; Sankaranarayanan, V.
Switched Finite Time Control of a Class of Underactuated Systems
99 p. 2006 [978-3-540-32799-8]

Vol. 332: Xu, S.; Lam, J.
Robust Control and Filtering of Singular Systems
234 p. 2006 [978-3-540-32797-4]

Vol. 331: Antsaklis, P.J.; Tabuada, P. (Eds.)
Networked Embedded Sensing and Control
367 p. 2006 [978-3-540-32794-3]